CONSTRUCTION IN CITIES

SOCIAL, ENVIRONMENTAL, POLITICAL, AND ECONOMIC CONCERNS

CONSTRUCTION IN CITIES

SOCIAL, ENVIRONMENTAL, POLITICAL, AND ECONOMIC CONCERNS

AUTHOR-EDITOR
Patricia J. Lancaster, AIA

CO-EDITORS
Edward S. Plotkin, P.E., FASCE
Jill N. Lerner, AIA

ADVISORY EDITOR
M.D. Morris, P.E.

CRC Press
Taylor & Francis Group
Boca Raton London New York

CRC Press is an imprint of the
Taylor & Francis Group, an **informa** business

Cover design by Marjory Dressler

CRC Press
Taylor & Francis Group
6000 Broken Sound Parkway NW, Suite 300
Boca Raton, FL 33487-2742

First issued in paperback 2019

ISBN-13: 978-0-8493-7486-9 (hbk)
ISBN-13: 978-0-367-39778-4 (pbk)
Library of Congress Card Number 00-045538

Library of Congress Cataloging-in-Publication Data

Construction in cities : social, environmental, political, and economic concerns / edited by
Patricia J. Lancaster with Edward S. Plotkin, Jill N. Lerner.
p. cm. — (Civil engineering-advisors)
Includes bibliographical references.
ISBN 0-8493-7486-3 (alk. paper)
1. Engineering—Management. 2. Civil engineering—Social aspects. 3. Civil. engineering—Environmental aspects. 4. Civil engineering—Economic aspects. 5. Civil engineering—Political aspects. I. Plotkin, Edward S. II. Lerner, Jill N. III. Title. IV. Series.

TA190.L35 2000
624′.09173′2—dc21 00-045538

Visit the Taylor & Francis Web site at
http://www.taylorandfrancis.com

and the CRC Press Web site at
http://www.crcpress.com

Editor

Patricia J. Lancaster is vice president in charge of public/private development for the northeast region at LCOR Incorporated. Prior to joining LCOR, Inc., Ms. Lancaster was assistant vice president for the Department of Planning, Design and Construction at Columbia University in the city of New York, where she headed up a five year $850 million capital program. Her responsibilities included long-term campus planning, new projects development, space utilization studies, and developing space rationalization and reorganization plans to accommodate immediate and future needs.

In 1994, Ms. Lancaster was named the first woman deputy commissioner for design and construction with the Department of General Services of New York City, where she managed a staff of 600 people with an annual construction budget of $500 million. Her 20 years of experience includes real estate development of Penn Station, the Foley Square Office Building, and a mixed-use building for Rutgers University with LCOR, Inc., as well as associations with top architectural firms, such as Perkins and Will and Michael Lynn, where she was associate director for interiors.

Ms. Lancaster is presently responsible for structuring and implementing public/private partnerships. She is currently working on financing, designing, building, and operating real property assets. She earned her Masters degree in architecture and is a licensed real estate broker and registered architect in the state of New York. She is a member of AIA, ADPSR, SCUP, AAPA, and a fellow of the New York Academy of Medicine, and a fellow of the Institute for Urban Design. In May, 2000 she was named one of the top 100 women in Real Estate by *"Real Estate Weekly."*

Contributors

Sherene Baugher, Ph.D.
Associate Professor
Department of Landscape Architecture
Cornell University
Ithaca, NY

Hillary Brown, AIA
Assistant Commissioner
New York City Department of Design and Construction
New York, NY

Jason Grabosky, Ph.D.
Assistant Professor
Department of Environmental Horticulture
University of Florida
Gainesville, FL

Joseph Guertin, P.E.
Senior Principal
GZA GeoEnvironmental, Inc.
Newton Falls, MA

John C. Hester, E.P., FICE
Operations Director
Balfour Beatty Major Projects
Surrey, England

Nicholas S. Ilijic, P.E.
Parsons Brinckerhoff
New York, NY
Former First Deputy Commissioner
New York City Department of Environmental Protection

Lenore Janis
President
Professional Women in Construction
New York, NY

Patricia J. Lancaster
Vice President for Public/Private Development
LCOR Incorporated
New York, NY

Deborah Leonard
Project Executive
Rockrose Development Corporation
New York, NY

Jill N. Lerner, AIA
Principal
Kohn Pedersen Fox Associates, PC
New York, NY

Francis X. McArdle
Managing Director
The General Contractors Association
New York, NY

Jon Marmor
Associate Editor
Columns Magazine
Seattle, Washington

Evelyn Mertens
Director of Public Relations
Professional Women in Construction
New York, NY

Edward S. Plotkin, P.E., FASCE
Consultant

Clive Pollard, C. Eng., FICE
Operations Director
Balfour Beatty Major Projects
Surrey, England

J. Patrick Powers, P.E.
Consultant
Mueser Rutledge
Marco Island, FL

Rueben Samuels, P.E., FASCE
Parsons Brinckerhoff
New York, NY

Richard W. Southwick, AIA
Managing Partner
Beyer Binder Bell
New York, NY

Gregory T. Waugh, AIA
Senior Associate
Kohn Pedersen Fox Associates, PC
New York, NY

Acknowledgments

On May 22, 1996 Lenore Janis took me to lunch at the Princeton Club in New York City, and introduced me to M.D. Morris. Lenore is the President of Professional Women in Construction and M.D. is the only man in construction to wear a brown tuxedo to formal events. M.D. asked me to write this book that day, and ever since has been a constant source of support and encouragement. Madelaine Stoller, Herb Bienstock, and Edith and Cyril Levett have been staunch allies during the entire process. Roni Gilbert has been my guiding light.

I would also like to thank my friends Elizabeth Hammond, Barbara Prowse Edwards, and Lorna Earl for their omnipresent belief in me.

Contents

Chapter 1 Not just bricks and mortar 1
Patricia J. Lancaster

Chapter 2 The challenges of constructing major tunnels in Central London 7
Clive Pollard and John C. Hester

Chapter 3 How things you can't see can cause problems 25
Joseph D. Guertin

Chapter 4 The contest with groundwater for underground space 49
J. Patrick Powers

Chapter 5 Mobilization for a tunnel project in an urban environment 59
Edward S. Plotkin

Chapter 6 Siting the North River wastewater treatment plant 69
Nicholas S. Ilijic

Chapter 7 Getting along with the existing infrastructure 77
Reuben Samuels

Chapter 8 Exterior wall renovation in urban areas 83
Gregory T. Waugh

Chapter 9 Community relations and urban design: the New York Psychiatric Institute case study 93
Jill N. Lerner

Chapter 10 Building under a city street 105
Deborah Leonard

Chapter 11 We've got an historic landmark, now what do we do? 117
Richard W. Southwick

Chapter 12 Turning archaeological problems into assets 135
Sherene Baugher

Chapter 13 Trees in urban construction 157
Jason Grabosky

Chapter 14 Building for high performance 193
Hillary Brown

Chapter 15 Construction labor in the urban setting 221
Francis X. McArdle

Chapter 16 Women in construction 231
Lenore Janis and Evelyn Mertens

Chapter 17 University of Washington case study 241
Jon Marmor

Appendix – The international perspective 245
Patricia J. Lancaster

Index 249

chapter one

Not just bricks and mortar

Patricia J. Lancaster

Contents

Environmental Regulation 2
Political Climate 3
Economic Issues 4
Social Legislation 5
This book 5

Your project is stopped! The community is outraged that you are not hiring locally. The press is following the politicians around as they claim you are creating an eyesore for their constituents. The building department seems to be slow in issuing permits for your particular project and no one knows why. An ancestral burial ground has been unearthed on your site right next to the leaking oil tank from the land-marked outhouse that you were going to demolish until the preservationists started screaming. And then an adorable whooping crane couple decides to nest in your de-watering zone. You sigh and decide to go around the corner for a cup of coffee only to discover that one of the community activists has put your photograph on a web page, and you are being mobbed by angry neighbors.

In the 1960's, you could have awakened and that all would have been a bad dream. Now, it is your life. You as the construction professional must not only be able, technically, to build the building, but must be able to get it built amid all sorts of adverse forces that have a power today that they have never had before. In some cases, the power is legislated and in some cases it is not, but in all cases it is capable of slowing or halting your project. Gone are the days when you can rip down a building in the middle of the night as Harry Maclowe did in New York in 1967; or use uranium trappings

0-8493-7486-3/01/$0.00+$.50

to make concrete as the federal government did in Colorado in 1971; or build plants that excrete toxic waste as Dow did at Love Canal.

Federal mandates enacted during the 1960s and 1970s for government projects had the net effect of empowering the public on all types of projects. Now there are regulations about everything. People believe that individually and collectively they have a right to know about and to influence the progress and outcome of any project. Jurisdictional borders are nebulous enough for the projects to be slowed or stopped even if they are being built "as-of-right" (without the need for any regulatory approvals). Design and construction now go hand in hand with policy, people, and process.

Environmental Regulation

Environmental policies were generated at the federal level and came about as a result of lobbying by large numbers of people concerned about saving the earth after the flower children and hippie movements in the 1960s. Democrats were in power. Nobody really thought that saving the earth was a bad thing, nor probably had any idea of the far reaching impact of the new laws. The construction industry did not really participate in their formulation and did not lobby either for or against. Here is a chronological summary of those regulations:

- 1963 Clean Air Act
- 1966 National Historic Preservation Act
- 1969 National Environmental Policy Act (NEPA)
- 1972 Clean Water Act
- 1973 Endangered Species Act
- 1976 Resource Conservation and Recovery Act
- Early 1990s Global Conference in Rio de Janeiro regarding comprehensive environmental assessment vs. performance.
- European Union Mandate

The concept of an interdisciplinary, comprehensive environmental impact assessment was first introduced when the Congress included it in Section 102(2)c of the National Environmental Policy Act of 1969 known as NEPA. NEPA regulations require all federal agencies to evaluate the environmental consequences of proposed actions and to consider alternatives. The concept of this mandate on federal projects led the public to believe that they had a right to participate in making decisions on projects, and whereas the policy only applied to government projects, the concept has had more far–reaching implications for the construction industry.

The major thrust of this law was to improve and plan for the natural environment. States then began to want jurisdiction and control also, and to vie with the federal government for who would adjudicate the new concepts. In 1975, for example, New York State's legislature enacted the State Environmental Quality Review Act known as SEQRA. This act requires that all state and local governmental agencies assess environmental effects of discretionary actions (unless such actions fall within specific statutory or regulatory exemptions from the requirements for review) before undertaking, funding, or approving the action.

By the late 1980s, the inclusion of environmental measures into contract specifications had developed into the necessity of monitoring and mitigation programs. The movement now is toward a more holistic approach initiated by contractors themselves in response to the market need. Such programs have become more common with long-term projects in which environmental issues cannot always be foreseen. Most of these programs now advocate evaluating a construction project from acquisition of raw materials through manufacture, construction, and eventual demolition with an eye to preventing or solving environmental problems before they start; and to use appropriate products and methods in the process.

The 1990s have given birth to a series of refinements, such as the establishment by NEPA of the President's Council on the Environment (that President Clinton tried to disband), and the revisions to the Clean Air Act that involve the size of particulate matter viewed as harmful.

NEPA and the other acts gave the public rights. In the next ten years we are unlikely to see major changes, only perhaps shifts in emphasis as more legal precedents are established. For instance, there are now people suing for poor indoor air quality under the Americans with Disabilities Act (ADA) which initially sought only to ensure dignified and safe access to those unable to perambulate well. Projects are now evaluated not only for their performance but for all their ramifications. The acts are evolving not as set standards, but as guardians of the public's right to full disclosure of all the facts, figures, and possible alternative solutions. Projects are more likely to be stopped by lack of disclosure than by any specific item disclosed.

Political Climate

It is important to know the political climate, and to understand the agendas and priorities of those in power, to get projects built. The design and construction industries are not well versed in how to achieve political gain, whether it be project-oriented or legislative. Out of the six hundred or so legislators in Washington, only about 5% are from our field. The people who exert influence, therefore, are the elected officials' constituents and those not involved with the construction industry. Of the constituents who vote for any given elected official, only about 10% are active in trying to influence the decisions of that official. This 10% is generally led and directed by 1% of the constituents, and the rest follow. This means that if you want a project approved, you need to know who the power players are. This obviously applies to projects that need public approval, such as ones that require Environmental Impact Statements (EISs) and have public hearings. Power players can also affect as-of-right projects where strong anti-project sentiment exists.

The design-end people of the profession have generally seen themselves as being above exerting political influence, thinking that if they state a good, rational case with good back-up information, the elected official will make a well-informed, rational decision. This is rarely true, and the sooner we realize it, the better off the profession and individual projects will be.

Look at the issue of the watershed for the New York City water supply. Upstate constituents wanted to develop the land around the reservoirs; that would have contaminated New York City's water supply and forced the city to spend $800 million purifying its water. If you were working on a house for one of those developers, did you know how to influence the fate of part of your livelihood? The number of people influenced was many; the number of power players was few.

Particular legislation will affect your particular life or livelihood. But also, the trends towards reviews, regulations, and legislation will have an impact on your ability to proceed. The profession has a new need to be aware of such trends, to know the effects they will have on our projects, and to gain the capacity to influence the law-makers. By the way, environmental concerns are issues that cross political party lines.

Economic Issues

The construction industry has a great impact on the economy but is barely recognized. A rough rule of thumb is that for every one construction dollar spent, eleven dollars in economic impact are generated. This turns into an enormous, enigmatic economic engine. Imagine the money spent by the designer, engineer, construction manager, contractor, sub-contractor, supplier, and fabricator on any given project. For the reconstruction of the LaGuardia Airport, for instance, the Port Authority of New York and New Jersey recorded $7 million in receipts for construction lunches, contractor supplies, office supplies, and local vendor purchases. We need to capitalize on our effects, and make people aware of our enormous buying power.

Construction is one of the only industries that can still generate blue–collar jobs, yet it is one of the most complex and diffuse in the economy. It has never perceived itself as a whole industry, but rather as a series of fragmented, self-important businesses that have a time-honored place in the creation of one of humanity's most basic needs: shelter. This probably goes back to the idea of master builder and the craft guilds, but now is a little outdated. As blue–collar manufacturing jobs are being lost to mechanization and technology all over the nation, construction is becoming one of the only middle class job options. It is also in need of young people to train and work. Yet the best and brightest young people are not choosing our industry. The industry has evolved over the years in some ways to be its own worst enemy. It is fragmented and fraught with many different kinds of rules and regulations. A big project can put thousands of people to work relatively quickly, and the physical structure that results is a legacy for the community in which it is built. The industry should try to insist that public policy demand both the re-establishment of the middle class, and restoration and enhancement of our physical assets and infrastructure.

Social Legislation

It is imperative that you know the importance of ad hoc community interest groups. Whether a large project or small, you can be hit by a special interest group. Right now, New York City has what is called fair share legislation. This law entitles not only those in the immediate project area but under served communities to their fair share of the economic and social benefits of every public dollar spent. This legislation is relatively new and may not seem particularly apropos for smaller communities, but it is indicative of a trend toward the necessity for due and just consideration of the larger public good. Of course, project delays or stoppages have economic in addition to social impacts.

This book

Each ensuing chapter will address aspects of the consensus crisis facing the construction industry, and show how diverse environmental, economic, political, and social agendas can shape, stop, and propel projects. As a construction professional, you must be adept at knowing the concerns likely to be raised during your project, and at formulating a strategy in advance for how to mitigate adverse non-construction issues. Our industry has seen projects live or die by the prospect of regulation and litigation. One common denominator in many of the projects that did not happen is the aspect of surprise, or lack of advance information. Owners and developers often lose interest in a project when there is a potential for many delays caused by the need to follow a long, cumbersome litigious or regulatory road. This is especially true in cases where the extent of mitigation was not clearly identified at the onset of the development process. On the other hand, many successful projects were developed on extremely difficult sites. The differentiating element in those cases was the presence of advance knowledge and understanding of the probable actions required, and the preparation of a proactive plan. What may seem like the fastest and cheapest method or sequence may not turn out that way. The specific actions taken on projects in these chapters and the concise delineation of issues and solutions given will equip you to better manage your projects.

chapter two

The challenges of constructing major tunnels in Central London

Clive Pollard and John C. Hester

Contents

An American's view 7
The beginning 8
The Parliamentary effect 10
Finding suitable locations 11
Archaeology 12
Limiting damage to the existing infrastructure 13
Unforeseen geological conditions 19
Protecting the environment 20
Material supply and disposal 22
Conclusion 24

An American's view

A Joint Venture of Balfour Beatty Major Projects and AMEC Civil Engineering was awarded the contract to build the Waterloo and Westminster section of the new Jubilee Line in late 1993. Major construction work commenced in the spring of 1994. Although the combined tunneling experience of these two companies is second to none in the United Kingdom and must be near the top of any worldwide rankings, there was no experience in the type of settlement control and restrictions required for this project. Possible damage to such structures as Big Ben, Waterloo station, and the surrounding infrastructure was to be avoided by the use of a relatively new technique called "compensation grouting." Although trials of this system had been

0-8493-7486-3/01/$0.00+$.50

carried out prior to its use in critical situations during actual construction, I think most tunneling people still regarded grouting as a "black art."

London is not a city that is continuously rebuilding itself as is the case with most cities in the USA. I think London very slowly evolves and therefore construction of major projects must not destroy the existing structures, but expand upon them and improve them to fit the new project. I don't think that the destruction and replacement of Waterloo Station, as happened to Pennsylvania Station in New York, was even considered. If Madison Square Garden had been in London, it is very likely that it would still exist in its original external appearance and have been modernized internally.

As detailed later in this chapter, London is built on clay. London clay is very dense and an excellent tunneling formation but circumstances often do not lend themselves to the use of earth pressure balance or slurry shield techniques. It is, therefore, almost impossible to eliminate settlement without some external system.

After several years of association with this project, I am still amazed that it has been built. It must have taken mountains of confidence in the engineering fraternity to persuade all of the relevant agencies and the insurance companies that the signature area of Great Britain would be preserved intact by grouting.

It did happen, and a partial explanation of the results and further background to this achievement follows.

J.C. Hester

The beginning

London has existed since Roman times and, until the nineteenth century, it was the largest city in the world. Today it is still one of the world's largest cities with a population of 7 to 9 million, depending on where you consider the limits of its sprawling, densely built-up urban area.

London has always been known for having an excellent tunneling medium beneath its streets. However, this thick layer of blue London clay occurs mainly to the north of the river Thames. To the south, there are deposits of sands and gravels in a wide variety of gradings, depths, and densities which until recent years were outside the limits of cost effective tunnel construction.

When the advantages of the London clay were discovered in the late 1800s, the underground railway system developed extensively north of the Thames. To the south, where ground conditions were not then favorable for tunneling, the transport system was developed as an extensive surface railway network, often built on brick viaducts, bridges, and embankments.

In the years following the second World War, the demise of London as one of the world's greatest ports left the large docklands area in the east of the city available for re-development. The viability of this area required new transport links. These included one running from west to east, generally south of the Thames and linking with the existing transport networks both north and south of the Thames. This link is known as the Jubilee Line Extension Project (JLEP) and is valued at £3 billion (1999) (Figure 2.1).

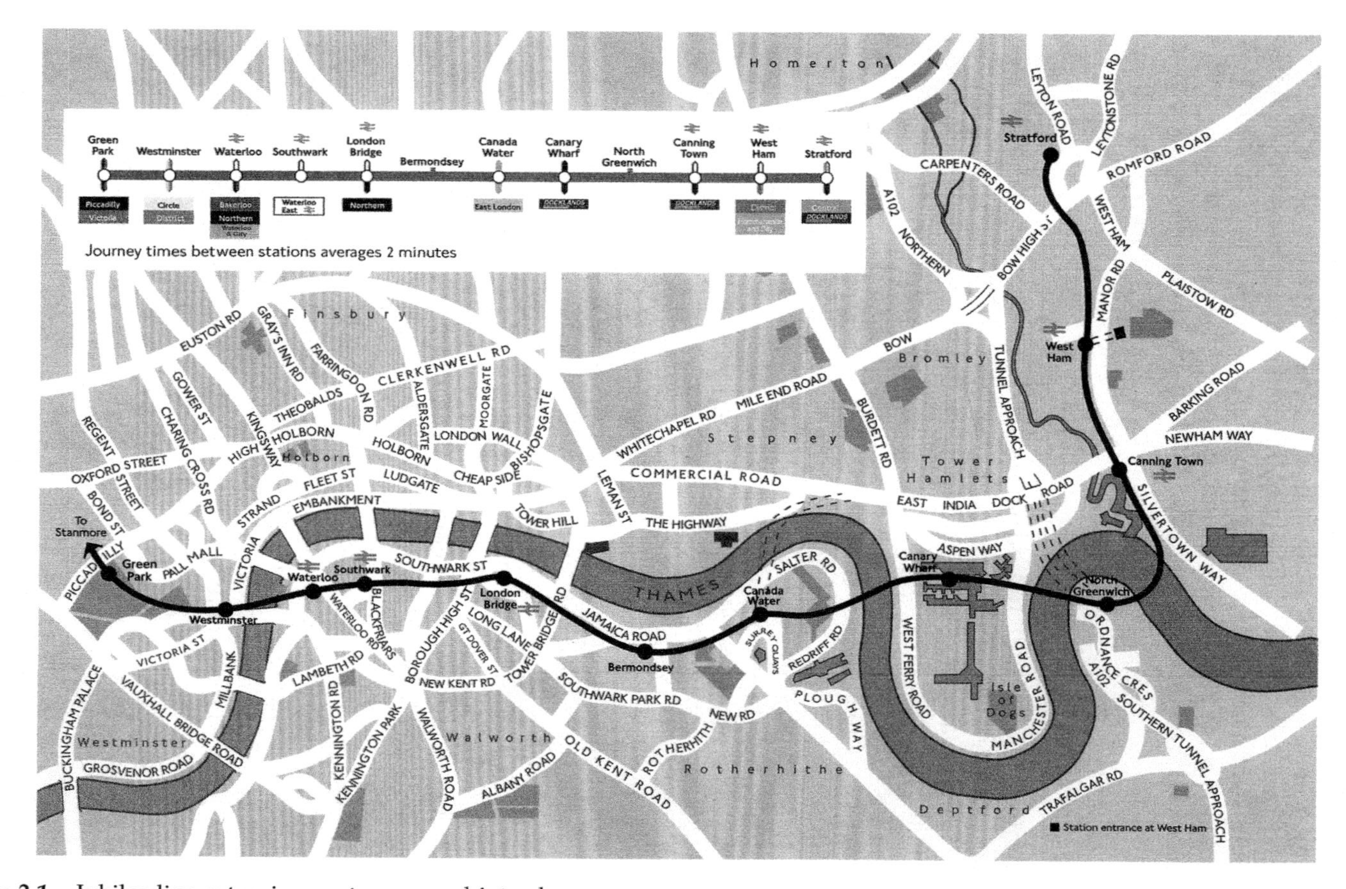

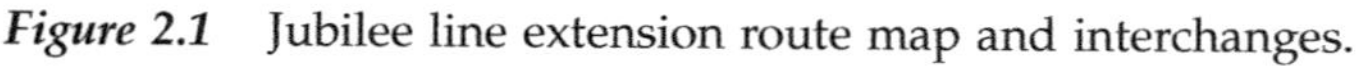
Figure 2.1 Jubilee line extension route map and interchanges.

The JLEP team carried out the design and management of the overall construction of the Jubilee Line Extension for London Underground Ltd. The construction of the Waterloo to Green Park section of the project was carried out by a joint venture of Balfour Beatty Major Projects and AMEC Civil Engineering. This contract, number 102, included new stations at both Waterloo and Westminster and was the largest on the project by a factor of two. It included some 20% of all the civil engineering work on the project. This project will be used to illustrate some of the challenges of tunneling in central London.

The Parliamentary effect

At the time that the JLEP was being developed, it was still necessary to pass an act of Parliament for any new railway in the U.K. This process meant that the scheme became very high profile and the way was opened for all manner of petitioners to express reservations and seek changes to the scheme to satisfy their own particular interests.

Typical of these were:

- Archaeologists
- The Royal parks
- Public utilities such as water, sewage, power, and a number of communications companies
- Public services such as fire fighting, ambulance, and three branches of the police force
- A number of separate operating zones of Railtrack which are responsible for surface railway infrastructure
- All local authorities along the route
- The London Traffic Directorate
- A variety of environmental and historical interest groups
- A large number of influential or prestigious, independent organizations and building owners along the route

The concerns of the groups covered such aspects as noise and vibration (both during construction and operation), settlement effects on surface structures, loss of amenities during construction, integration of the works into the existing amenities, and environmental effects.

The Jubilee Line required three Acts of Parliament before all parties were satisfied with the arrangements and the process took almost two years to complete. As a result, London Underground Ltd. (LUL) entered into something of the order of 190 undertakings with a variety of parties as part of the parliamentary process. Of these, 55 were in the Waterloo to Green Park section.

In addition, LUL entered into a large number of other special written agreements with third parties to govern rights in relation to the design, construction, and operation of the railway.

In addition to the parliamentary process, Parliament has many other rights which have developed over many years and which affect construction sites in their vicinity. For example, voting at the end of parliamentary debates — when "the house divides" — is in person. It is also a fact that most debates take place in the afternoons and evenings. One of the MP's rights is that they must not be obstructed on their way to a "division." If the work requires night-time road closures, these cannot be put in place until Parliament has finished sitting for the night. If the sitting runs late - so does the road closure!

In a similar manner, works that are required to be carried out within the Palace of Westminster can only be carried out in periods when "the house" is not sitting.

This area of Westminster is a focus for a number of ceremonial occasions. It is also a focus for demonstrations, by all sectors of the community, who seek to influence Parliament. These can often require stringent security measures, which also affect construction activities in the vicinity. Also, as a major tourist attraction, it must be one of the few areas in the world where a major traffic management consideration is the width of the footpath to accommodate the people!

Finding suitable locations

A fundamental problem in any city center is finding space for the access points to any new underground facilities and then to locate them such that they have minimal effect on the existing surface and sub-surface structures. In London, the obvious solution of demolishing existing structures is rarely acceptable.

Where there is London clay there will, almost certainly, already be tunnels for underground lines and other sub-surface utilities. Thus at Waterloo, there are three adjacent surface railway stations. Waterloo International is a very modern station where trains leave for Paris and Brussels via the Channel Tunnel. Waterloo mainline station serves the railways to the south and west of London and Waterloo Junction links to Charing Cross just north of the Thames and London Bridge station to the east together with East Kent and the Channel ports. Beneath this surface complex there are twin tunnels for each of the LUL Northern, Bakerloo and Waterloo and City Lines. For the new Jubilee Line, it was necessary to insert into this space an additional two platform tunnels plus the necessary escalator, lift, and ventilation shafts to create one of the most important stations on the Jubilee Line (Figure 2.2).

At Westminster, the problem of space was equally acute but for totally different reasons. Here there are the Houses of Parliament with St. Stephen's Clock Tower (Big Ben) – one of the most famous landmarks in the world – Westminster Abbey, Westminster Bridge and the adjacent river walls. The existing LUL Westminster station is quite shallow, having been built by cut and cover methods when steam trains were in use. Next to the station is an 8-foot diameter brick built sewer, built in about 1870, serving about 20% of the population of London. Nearby is a 30-inch diameter high-pressure water main also serving a large sector of London.

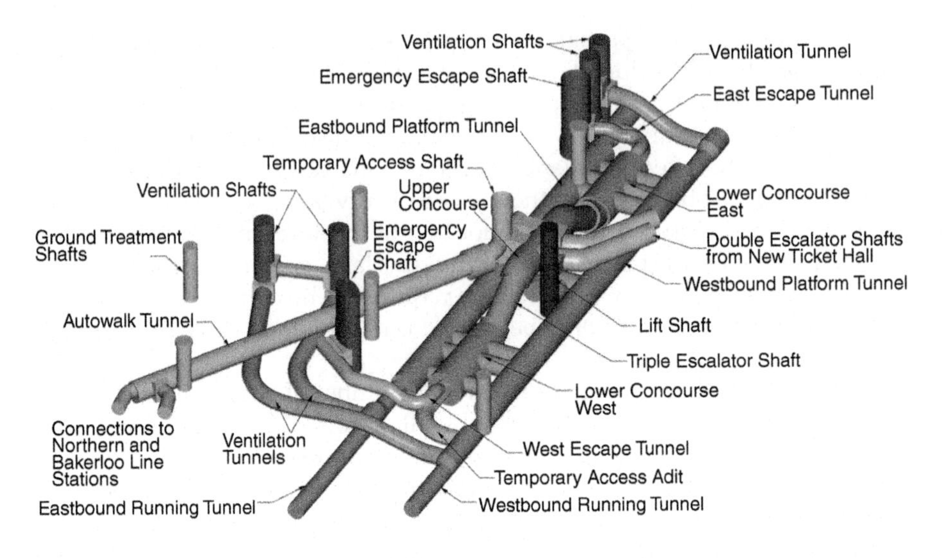

Figure 2.2 Isometric view of tunnels and shafts of Waterloo.

The site that was initially chosen for the Westminster Jubilee Line station was under Parliament Square with a pedestrian subway link to the existing District and Circle Line (D&CL) station. However, the various interested parties did not accept this location.

After much debate, the site that was finally chosen was immediately below the existing D&CL station. This was also to be the site of a new, prestigious building to provide offices for the Members of Parliament and the new station structure had to be designed to incorporate the foundations for the new building. This required the existing D&CL station to be rebuilt and lowered 300mm to meet the architectural requirements of the new building. A 12-story deep basement structure was then required to accommodate all of the access escalators and lifts to service the twin platform tunnels and associated access and ventilation tunnels. These tunnels had to be stacked one above the other because of the proximity of Big Ben. All of this complex arrangement was required to be built with only limited weekend closures of the D&CL station and according to a timetable agreed with Parliament (Figure 2.3).

Archaeology

London has a history going back more than 2000 years and excavation through the superficial deposits is seen as a major opportunity for the archaeologists to discover more about the past. Archaeological requirements were therefore built into the Jubilee Line construction contracts with requirements to allow specific periods of time for archaeological investigation in the construction programs.

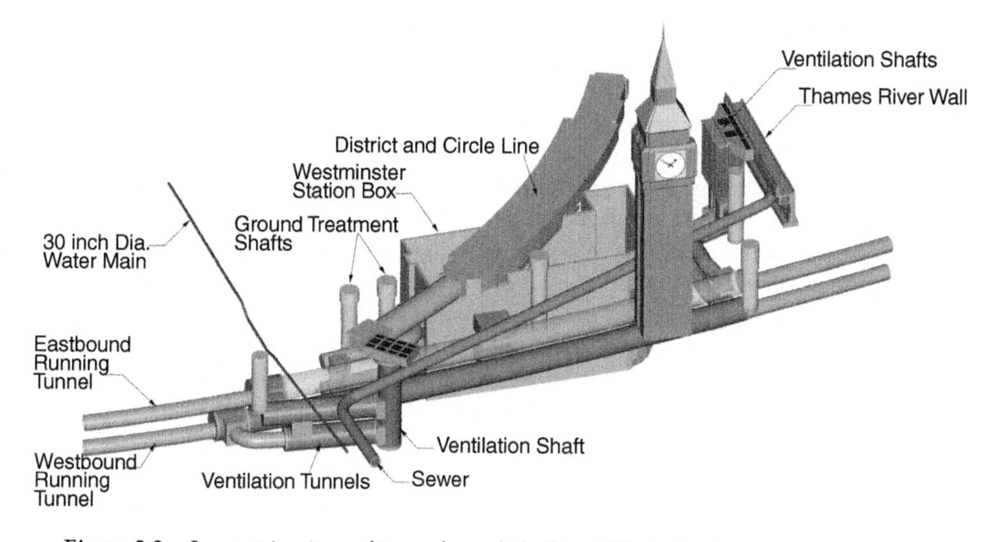

Figure 2.3 Isometric view of tunnels and shafts at Westminster.

On occasions, the period allowed would be more than sufficient and on others, the discoveries would mean that more time was required. On others, the discovery would be a total surprise such as the stones from an old river wall and a child's skull found when sinking a shaft at Westminster. The actual archaeological excavation work was carried out by the Museum of London's archaeology service. Along the whole length of the line this proved to be a very fruitful exercise.

At Westminster, a Neolithic (5000 to 2000 BC) flint arrowhead was found together with a fragment of a stone axe and other flint tools. There were also fragments of late Bronze Age or early Iron Age (700 BC) pottery and fragments of Roman building materials. In fact, there was evidence of continuous human occupation of this area up to the present day. The finds included two complete ceramic watering cans from the 16th century and an intact Delftware jug dated 1627 which was found in an old well.

Further east, at London Bridge station, they uncovered a wealth of information about London's Roman period and, at Stratford, the excavations uncovered the Cistercian, Langthorne Abbey, and the remains of its surrounding community. This abbey was founded in 1135 and was closed in 1538 by Henry VIII's dissolution of the monasteries.

The huge extent of these archaeological discoveries puts into perspective the fact that modern construction is just another episode in the life of a great city.

Limiting damage to the existing infrastructure

While archaeology is seen as an essential part of the workings in London, in construction terms it is a relatively minor factor. Much more important and more difficult to deal with are the effects upon existing structures. These are often of national importance and have been constructed many years ago to standards that did not anticipate differential settlement from deep excavations either directly beneath them or close by. Damage to such structures had to be kept to a minimum.

During tunneling, there is a tendency for the exposed areas of the excavation to move into the tunnel. The extent of this movement is dependent upon the ground conditions and the tunneling method adopted. The result is a settlement trough above the tunnels. The basis for calculating this settlement is usually expressed as a percentage of the face area and is called "face loss." While in some circumstances tunneling machinery can be designed to minimize this face loss effect, this opportunity is extremely limited with the very short drive lengths which occur in the construction of underground railway stations.

The first step in any assessment is to calculate the possible settlement contours from the tunnel layout using an assumed face loss and assuming a "green field site" i.e., no buildings or other structures in place. Using these contours, the effects on the buildings within the settlement trough can be assessed thus revealing those buildings or other structures which are at risk from the effects of the tunneling. At Waterloo and Westminster, the figure used for this engineering assessment was a 2% face loss.

The assessments for the new stations at Waterloo and Westminster showed that the settlement in some areas would cause unacceptable damage to the existing structures. It was therefore decided that the new technique of compensation grouting would be employed and this relatively new technique was incorporated in the construction contracts.

Some examples of these structures and services were the 8 feet in diameter, brick built, trunk sewer at the Westminster, Waterloo station with its numerous escalators and the brick arch railway viaduct to Charing Cross. Following a contract award, Big Ben was added to this list.

Theoretically, compensation grouting is the injection of carefully calculated volumes of grout into the ground between the tunnel and the structure such that it compensates for the "face loss" and prevents the settlement effect from actually reaching the surface.

Compensation grout is usually injected into the ground, "tubes à manchette," which have been drilled in a horizontal array from a nearby shaft. Therefore having found the buildings that are at risk, one now has to find locations for the shafts and the arrays to enable the grout to be injected as the tunneling proceeds. In the Westminster area, this required an additional 11 shafts and, at Waterloo, an additional 6 shafts plus the use of two of the permanent shafts. Finding locations for these shafts and their associated equipment was not easy. Shafts were sunk from the basements of shops and banks, in the middle of busy streets with all the necessary traffic diversions, in private gardens, and within the labyrinth of brick arches that support Waterloo station.

A simple example of the problems encountered was that of the fine art dealer who pointed out that if a shaft site was established outside of his shop, he would be unable to get his larger pictures through the entrance to his premises.

From the shafts that were eventually established, nearly 38 Kms. of 75 mm-diameter "tubes à manchette" were drilled to produce a grid of injection points in the clay beneath the structures which were at risk (Figures 2.4 & 2.5).

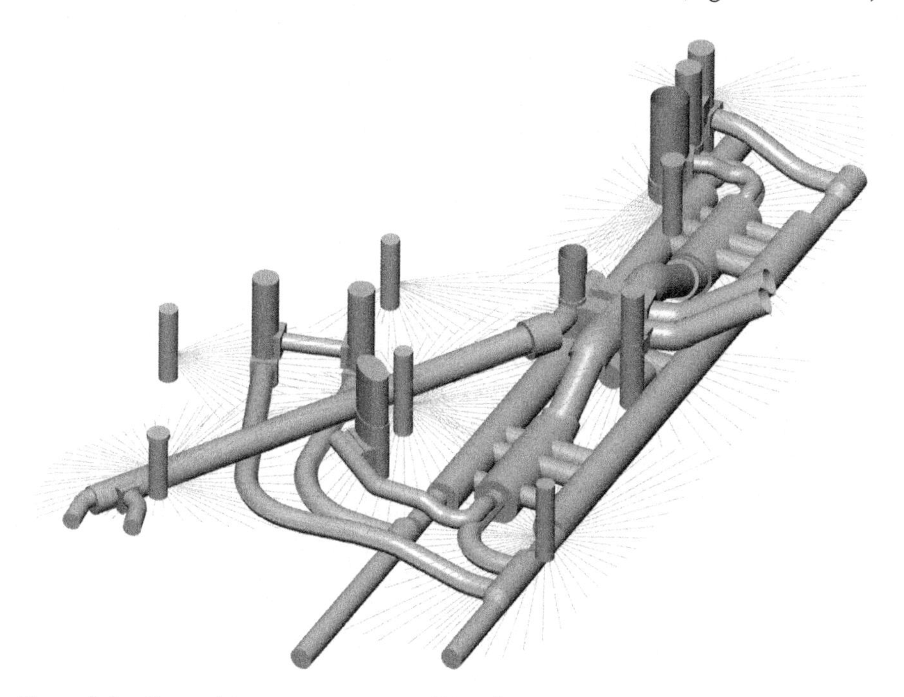

Figure 2.4 Ground treatment arrays at Waterloo.

A vital aspect of the compensation grouting process is that of monitoring ground and building movements while the tunneling and grouting are taking place. For the Waterloo and Westminster stations and associated tunnels, this required a comprehensive system which had the capability of producing useful information on a real time basis. A computer system was set up on site and software developed to handle the data. This data was received from data loggers via 37 telephone modems and information from field surveyors downloaded to the file servers via mobile telephones. The data related to more than 10,000 leveling points, 300 electro-levels, and a variety of other instrumentation such as inclinometers and extensometers. Many of these were in buildings or on operating railways and again, the consent of the owners was required to allow them to be fixed and maintained.

From this data, time–displacement plots and displacement contour plots could be produced at various time intervals. These were reviewed either

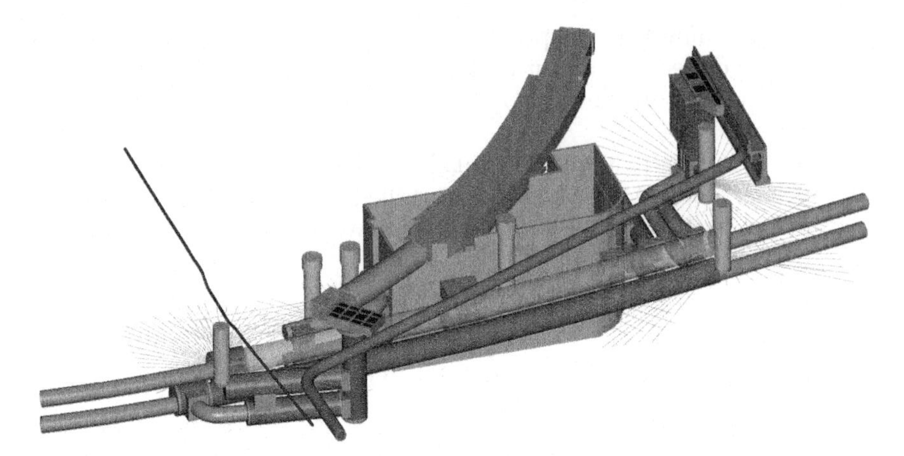

Figure 2.5 Isometric view showing ground treatment arrays at Westminster.

daily or at intervals to suit the excavation process. On occasion, the excavation process had to be interrupted to allow the compensation grouting to take place. The results were compared to the anticipated results and further grout injection patterns were modified to suit.

In the Westminster area, the main concern was Big Ben. To monitor its position, a 55 m long plumb line was hung inside the tower over a digitiser plate that recorded its position at three-minute intervals. Two grout arrays were established under the tower. The results of the grouting and monitoring are shown in Figure 2.6. This also shows the effects of weather and tides on the tower. It will be seen from the graph that the movements were contained to about 27 mm horizontal movement at a height of 55 m – just fractions of millimeters in terms of differential ground levels at the base of the tower.

In the St. James area, a structure of a very different nature had to be dealt with. This was a very ornate swimming pool in the basement of the RAC Club. This is an establishment with many very influential members who had advised all concerned of the dire consequences which would follow if their pool were cracked. There were just 7 m between the bottom of this pool and the crown of the tunnel and the predicted settlement of 20 mm was sufficient to crack the pool. A single shaft was established in the gardens of the club, and monitoring of the pool itself required the use of divers. A trial was carried out as the tunnel approached the area so that the necessary local ground parameters could be established. The net result was that the settlement was controlled to be less than 1mm, a remarkable achievement.

Compensation grouting is sometimes not the answer to settlement problems. In the particular case of the brick built sewer, a considerable length was found to require strengthening prior to allowing compensation grouting to be carried out in the vicinity. The engineering also becomes very complex when considering piled foundations.

In one area of Waterloo, the design of the station resulted in more than 60% of the ground being removed from beneath the foundations of the

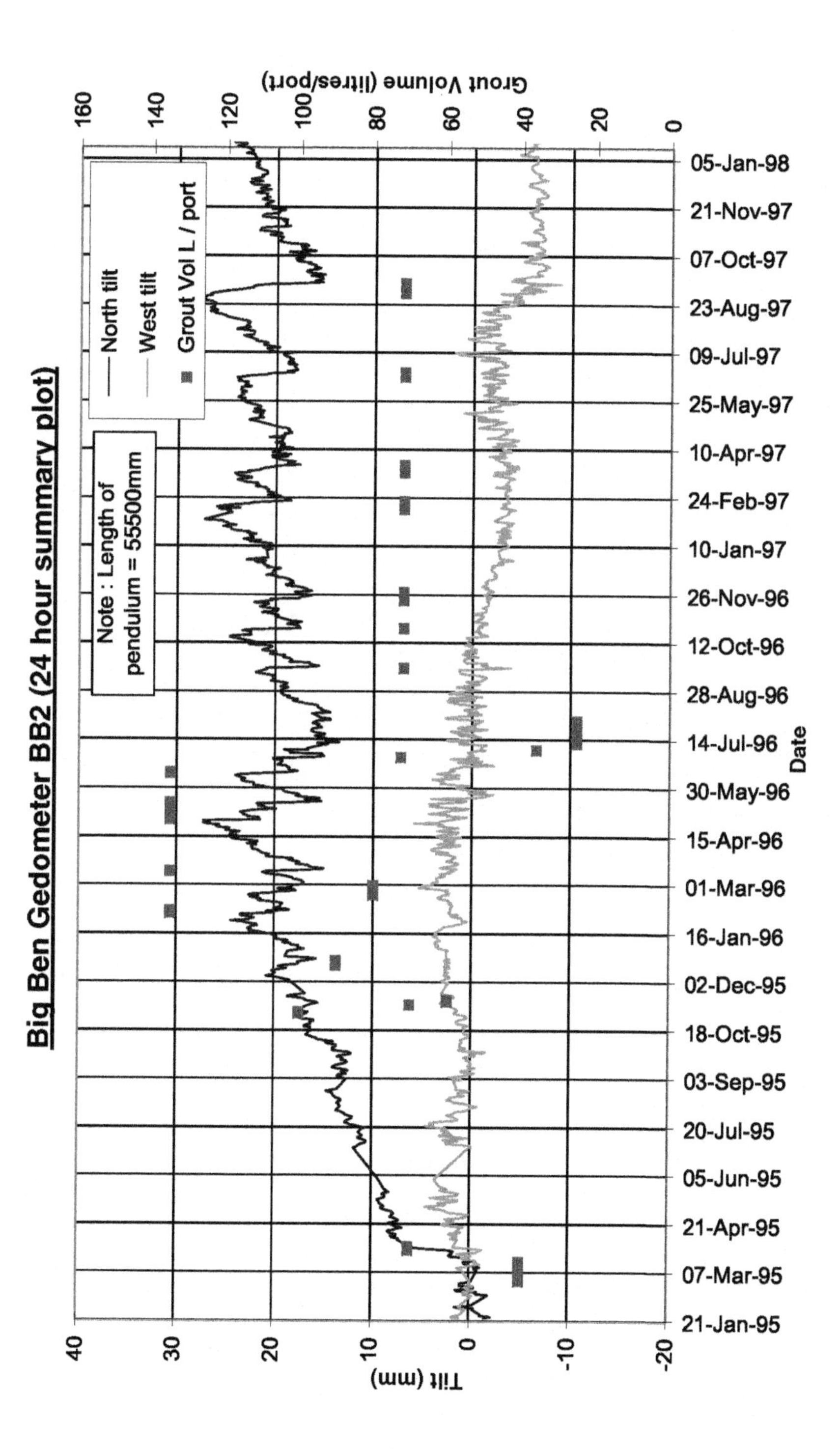

Figure 2.6 Big Ben Gedometer BB2 (24 hour summary plot).

railway bridge and the associated viaduct which carried millions of commuters per day in and out of Charing Cross Station. Settlements of more than 200 mm were predicted in this area. The tunnel construction took place at two levels. The lower level tunnels were constructed using sprayed concrete linings and consisted of a 6.7m diameter tunnel linking two 7m diameter platform tunnels, a 9m diameter lift shaft and its associated hydraulic pump station. At the upper levels, the tunnels were constructed using spheroidal graphite iron (SGI) segments. The main 10m diameter tunnel formed a concourse area where passengers using the pair of twin escalators from the surface and the Autowalk from the Bakerloo and Northern Line underground stations could gain access to the two triple escalator shafts leading down to the Jubilee Line platforms below. There was also a second junction to the lift shaft at this level (Figures 2.7 and 2.8).

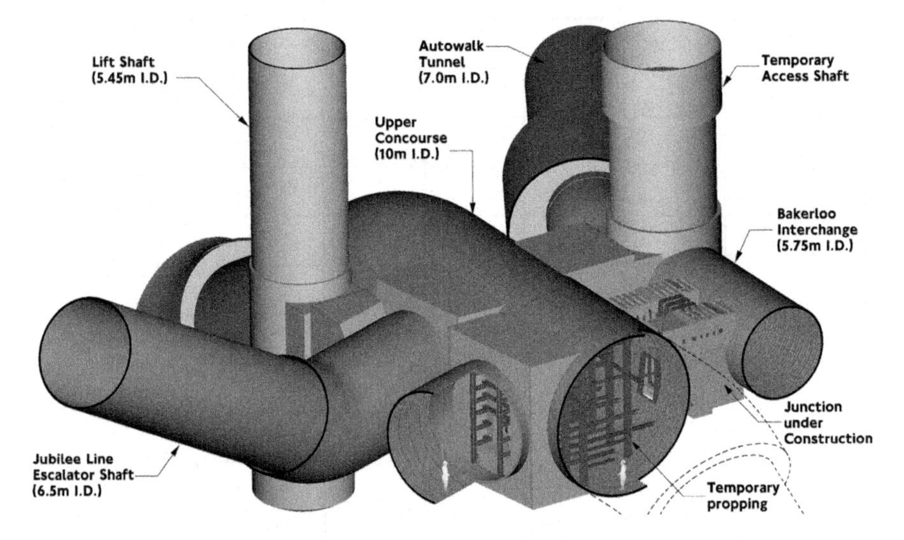

Figure 2.7 The complex junction of cast iron lined tunnels at the upper concourse at Waterloo.

The tunneling in this area was extremely difficult without the added complication of compensation grouting. The lift shaft was one of five shafts used for this particular section of compensation grouting. As with most shafts in London, it had been sunk through the Thames gravels as a caisson. This process can itself lead to settlement if ground losses occur. To guard against this, the gravels were permeated with silicate grouts in order to stabilize them. The area of Waterloo has been used for a multitude of industries in the past and, during drilling for the tubes à manchette, substances were encountered within the gravel which caused the drillers to become nauseous. Despite considerable research, the exact cause was never established and the work was completed using suitable personnel protection. The large upper tunnels were relatively close to the surface and the grout arrays, which were just in

Figure 2.8 Interior tunnel construction at upper concourse, Waterloo.

the clay below the gravels, were therefore also close to the tunnels. In this situation, the reactive loads from the grouting onto the tunnel created a need for additional temporary supports. Instrumentation revealed that the main steel props had become fully loaded at 230 tons each and further strengthening and careful phasing of the tunneling and grouting operations were required to avoid face collapse and possible catastrophic settlement above. In the event, with careful orchestration of the excavation and grouting, the works in this area were completed without any significant damage to the structures above.

Unforeseen geological conditions

The geology of the London basin is relatively straightforward with gravels overlying clay overlying further beds of sands. A real risk associated with tunneling in London is that of buried channels in the clay at the interface of the clay and the superficial water bearing gravels. Old streams formed these channels in the period before the gravels were deposited and are very narrow and difficult to detect in a dense urban area. During the construction of the Victoria Line in the mid 1960's, these were encountered on at least three occasions. The result is usually inundation of the tunnel and its equipment with free flowing sands and gravels and a hole breaking to the surface. Clearly this was an unacceptable risk in areas such as Waterloo and Westminster.

To guard against this risk, in all areas where the clay cover to the tunnels was less than 6m, the gravels were to be permeated with silicate grout for a minimum thickness of 3m. While this was possible in Waterloo, in Westminster the position of the brick lined sewer and other services coupled with a difficult traffic situation rendered the process almost impossible.

To overcome this problem, a "pipe arch" was constructed above the stacked platform tunnels. This consisted of five 1m diameter, concrete filled, steel tubes driven in both directions from a central chamber, parallel to and above the tunnels (Figures 2.9 & 2.10). These, as well as providing further security against the settlement of Big Ben, just 20m away, also provided ground information along the full length of the station and confirmed the absence of any narrow stream channels at the clay and gravel interface.

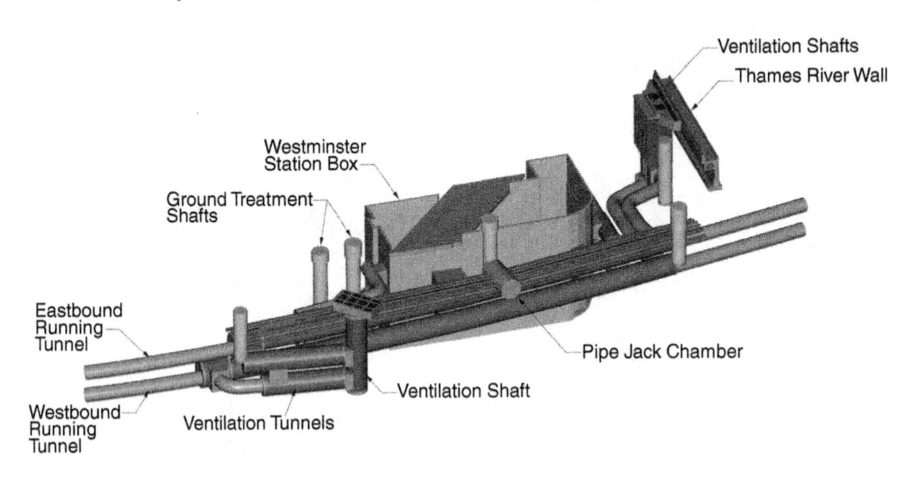

Figure 2.9 Isometric view of tunnels and shafts at Westminster "pipe arch."

Protecting the environment

The effects of man's activities on the environment have become very important in recent years. This has led to an increase in public awareness and an increase in legislation to protect the environment. It was therefore the policy of the JLEP team to adopt the most up-to-date standards.

As required by current U.K. legislation, a full Environmental Impact Study was carried out. There was a formal Environmental Policy and throughout the construction period, audits of the environmental performance of the various contractors were carried out. The results of these audits were published in order to obtain continuous improvement and awards were made to those contractors with the best performance.

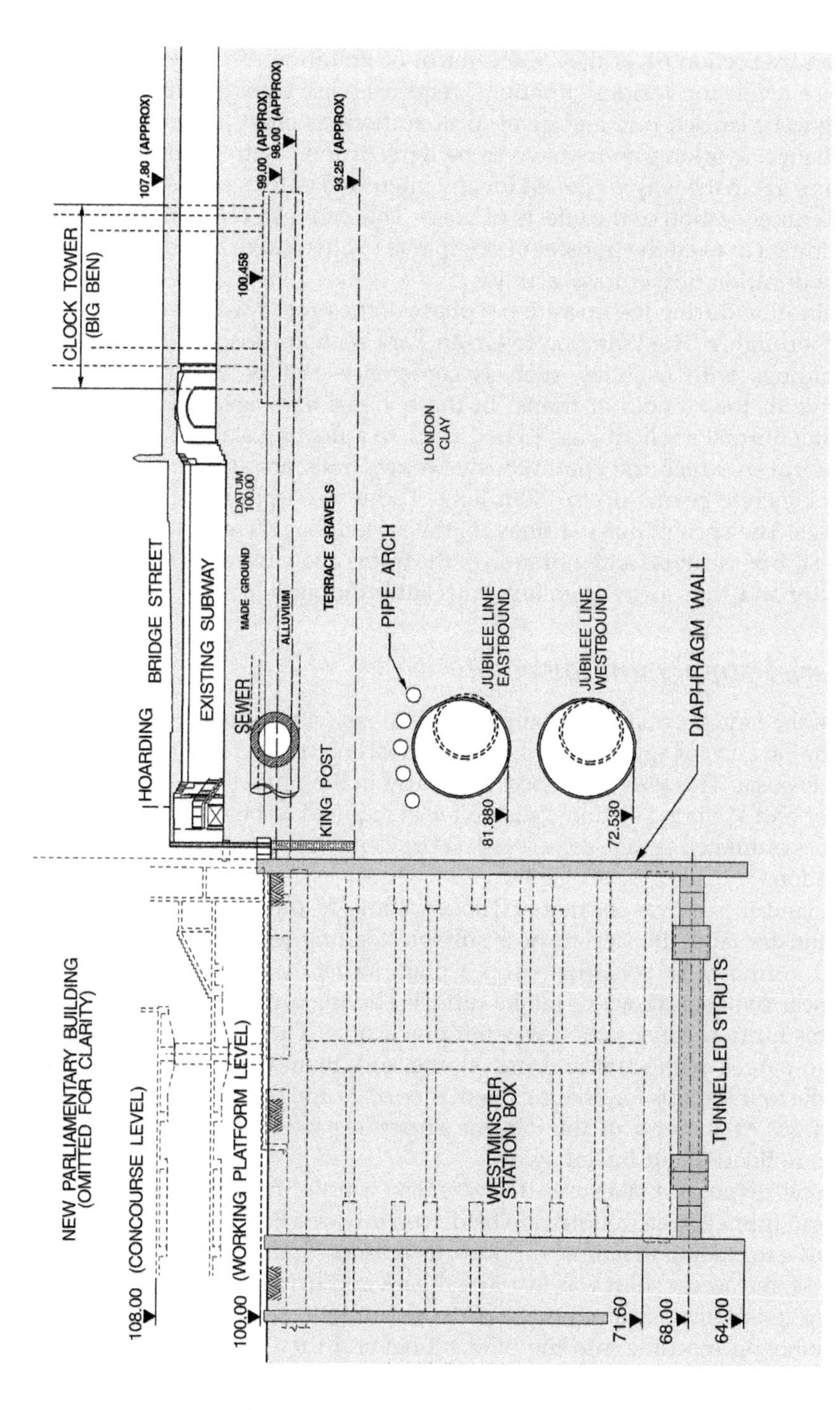

Figure 2.10 Section through Westminster Station box and tunnels.

Noise and vibration are key environmental aspects in urban areas and the JLEP team made a specific study of them. In the U.K., noise pollution is covered by Section 61 of the 1974 Control of Pollution Act. This sets limits on noise levels for various situations, requires noise assessments and monitoring to be carried out, and gives local authorities great powers to ensure compliance. Working hours have to be agreed upon with local authorities and once again the way is opened for any interested parties to raise objections or seek amelioration of the effects of noise. This can result in double-glazing to buildings or even the transfer of occupants to alternative accommodations for the duration of the noisy activity.

Vibration during the operational phase of the project was also dealt with very thoroughly. The Waterloo to Green Park section runs under a number of buildings with facilities, such as conference centers, which would be sensitive to the rumble of trains. In these areas, the tracks were changed from the normal resilient base-plated track to a floating slab track. This is a track form in which the continuously welded rails are supported on very heavy concrete beams up to 100m long. These are supported upon rubber bearings. The spring/mass stiffness of the system has the effect of reflecting about 80% of the noise and vibration of the trains away from the surrounding structure and thus away from any adjacent buildings.

Material supply and disposal

One of the largest environmental problems in any city is that of the effects of road traffic and for underground construction one of the largest problems is spoil disposal. This was addressed very early in the project and for the Westminster and Waterloo stations, all spoil was required to be removed by river. This was estimated to save at least 100,000 truck movements through the streets of London.

A landfill site was located at Tilbury, some 25 miles downstream from Westminster, and the design of a suitable loading jetty was started. This would normally be constructed on a piled structure but the location was very near the cooling water intake tunnels serving a nearby office building and this form of construction was not practicable. The neat solution was to fill two barges with sand and gravel and sink them on to the riverbed to form the foundations for floating boom moorings and the overhead conveyor discharge. At the end of the job, the structures were dismantled and the barges re-floated and towed away.

Spoil disposal at Westminster was also a little unusual. Here the two platform tunnels were required to be driven towards Big Ben. The excavated spoil in a tunneling machine, for obvious reasons, comes out of the back. In this case, the access shaft was in a Royal park and truck movement and spoil storage constraints required the spoil to be brought out through the front of the tunneling machine into the pilot tunnel and then to the main working site (Figure 2.11).

Figure 2.11 Westminster Station Shield showing a pilot tunnel dismantler and spoil disposal through the front.

At Waterloo, the available working site was too small to accommodate both segment storage and spoil handling facilities to meet an accelerated program requirement. In this case, the lower tunnels were constructed first and a "glory hole" constructed between the upper tunnels and these lower tunnels so that spoil could be dropped through it and transported underground to the main working site.

On all sites, the shortage of storage space meant that all deliveries of materials were on a "just in time" basis. This was particularly evident at the Westminster site where concrete reinforcement was delivered in quantities just sufficient for one shift's work and the trucks had to be loaded in exactly the right order for installation.

Conclusion

Tunneling in the center of any major city has its challenges but there are very few instances which cover the wide range to be found in central London. The success in overcoming these challenges, using a combination of tunneling techniques ranging from those well tried over many decades to the most modern of tunnel boring machines, coupled with state-of-the-art grouting, computer processing, and imaginative engineering, all in an environment with more than 2000 years of history, is a tribute to all those involved.

The author would like to thank the Jubilee Line Extension Project team and Balfour Beatty AMEC JV for their assistance and their permission to publish this account of the challenges of constructing major tunnels in central London.

chapter three

How things you can't see can cause problems

Joseph D. Guertin

Contents

An overview 25
The importance of geological setting 28
An historical perspective 28
An example of an earlier time – filling of Back Bay, Boston, Massachusetts 28
Tunneling then and now 30
Buildings and infrastructure; then and now 36
Excavation challenges 36
Underpinning challenges 43
The interrelationship of geotechnical and environmental issues 46
Summary 47
References..48

An overview

Construction in an urban environment has never been easy, but it was simpler during the first half of the 20th Century than after World War II. Urban construction is complex. Figure 3.1, which is a photograph of a small portion of the most complex urban highway construction project ever undertaken in the United States, illustrates the point.

Since the end of the war, intense public scrutiny has lead to aggressive political responses resulting in an ever-increasing body of complex zoning, design, and environmental regulations. At the heart of many of these regulations are geotechnical and environmental issues. Geotechnical issues

0-8493-7486-3/01/$0.00+$.50

Figure 3.1 Central Artery/Tunnel Project, Boston, Massachusetts.

concern the character and construction behavior of subsurface materials, i.e., soil, bedrock, and groundwater, whereas environmental issues cover a range of topics including the chemical nature of subsurface materials, air quality, construction noise, handling and disposal of construction waste, etc. A complete issues list would be very long. For the purposes of this chapter, the discussion of environmental issues is confined to soil, rock and groundwater contamination, and the management of these materials during and after construction.

Constructed facilities include roads, bridges, tunnels, buried pipes and conduits, etc., i.e., infrastructure, plus public or private buildings of all types. It might appear that the formidable technical challenges of underground and high-rise construction would surely control the process, but this is frequently not the case. While significant, these technical issues are just part of a long list of considerations with which project proponents, designers, and constructors must deal with to shepherd projects from vision to reality. Nontechnical considerations such as political issues, public scrutiny, environmental regulation, urban zoning, and design codes are just as likely to stop a project or be the cause for major changes as are technical challenges. This chapter presents a "broad-brush" description of the impacts of subsurface technical (geotechnology) and environmental issues and how they influence planning, design, and construction of new facilities in urban settings.

Geotechnology is that body of science (geology) and engineering (civil engineering) that is related to the description and behavior of soil, rock, and groundwater when altered by humans for the creation of constructed facilities, i.e. infrastructure and buildings. The civil engineering specialty concerned with the rational evaluation of ground and groundwater behavior as a result of civilization's desire to build things that rest on or are constructed in the ground is called geotechnical engineering. For example, design and construction of support systems to protect buildings, streets, and buried utilities immediately adjacent to deep foundation excavations are geotechnical engineering issues. Design and construction of building foundations, e.g., driven piles, are geotechnical engineering issues as well.

In a very broad interpretation, environmental issues include soil, rock and groundwater quality, air quality, traffic density, noise levels, light levels, worker safety, wildlife habitat (even in densely populated urban areas). For example, site contamination due to dumped or buried waste products will, in all likelihood, have to be characterized in detail and cleaned up, i.e., remediated, in accordance with specific regulatory requirements.

Geotechnical and ground-related environmental issues are frequently so closely interrelated that they are collectively referred to as "geoenvironmental" issues. This combining of technical disciplines is a result of the inevitable relationship between environmental quality and engineering behavior. In most cases, these issues must be dealt with in a coordinated way. For instance, if excavation is required for any reason, i.e., a building foundation, a tunnel, etc., and if the soil, rock, or groundwater is contaminated, the health risk to workers and proper disposal or containment of the contaminated material must be considered. These environmental and geotechnical issues are therefore dealt with in a coordinated, cost-efficient way.

The importance of geologic setting

Depending on the geologic setting, geotechnical issues can be of greater importance in certain cities than in others. For example, geotechnical considerations are often very important in coastal locations because of the greater likelihood that soft compressible soils will underlie these cities than those in mountainous regions. Large areas of cities such as Boston, Chicago, and San Francisco are underlain by soft soils; whereas cites such as Kansas City, Denver, and Phoenix are not. It is not surprising that pre-eminent universities where seminal geotechnical research is done are Boston (M.I.T. and Harvard), Chicago (University of Illinois) and San Francisco (The University of California at Berkeley).

An historical perspective

The significance of the environmental movement that resulted in the creation of today's regulatory system cannot be overstated. Prior to World War II and for ten to twenty years thereafter, environmental issues were of little concern to construction project planners and designers. People just didn't worry about it. Growth for growth's sake was good. Wetlands were filled to provide developable land, waste was dumped wherever it wouldn't be seen, and wildlife habitats were destroyed, all in the name of post-war economic growth. This attitude continued pretty much unabated until the late-1960's and 1970's when environmental concerns began to be taken seriously. The Federal Environmental Protection Agency was created in 1970. Similar state agencies were created at about the same time and the growth of regulation really began.

An example of an earlier time – filling of Back Bay, Boston, Massachusetts

Examples of the pre-regulatory climate where geotechnical issues were key to the growth of cities, but environmental concerns were not, are numerous. Many coastal cites such as New York and Boston were enlarged through filling of tidal flats. This wholesale filling of wetland areas would not be allowed under current environmental law. One of the most striking examples that led to the creation of major sections of Boston, as we know it today, was the filling of Back Bay. Approximately 600 acres of tidal mud flats were filled over a 137 year period from 1814 to 1951, however, most of the filling took place between 1814 and 1892. Figure 3.2[1] shows the Back Bay in the late 1850's and Figure 3.3 illustrates the history of filling operations from 1814 to about 1871.[2]

The work was started as a means to create more useable land area to the west of the original colonial Boston peninsula and to fill areas that had become foul smelling mud flats from city sewage. The fill was placed over

Figure 3.2 Back Bay in Boston, Massachusetts in the late 1850's.

soft marine sediments and clay with subsequent area settlement, but was accomplished successfully. Nearly ten million cubic yards of granular fill were imported over a 10-mile long specially constructed railroad between the suburb of Needham, Massachusetts and Back Bay. Figure 3.4[3] is a photograph of the borrow pit operation where sand and gravel were obtained for fill.

Public buildings and homes were constructed on the newly filled land resulting in what today is the fashionable Back Bay section of Boston.

The Charles River tidal estuary located immediately north of Back Bay had long been an aesthetic issue for the city. Twice a day at low tide, the resulting mud flats presented health problems from sewage being discharged into the Basin and the general unsightliness of extensive mud flats behind some of Boston's most fashionable 19th century homes. As early as 1859, the idea of building a dam across the mouth of the Charles River to create a fresh water lagoon that would cover the mud flats, thereby improving sanitary conditions and creating a body of water for public use and enjoyment was discussed. Twice a day, homes overlooking the Charles viewed dreary foul smelling tidal flats. In the spring of 1902, Charles R. Freeman accepted the position as Chief Engineer for the Committee on the Charles River Dam. In a report submitted only nine months later, the Committee recommended the construction of a dam, and in 1910 the dam was completed. Figure 3.5 is a photograph of the Charles River Basin approximately 100 years after it was created. The wisdom of these early engineers is obvious.

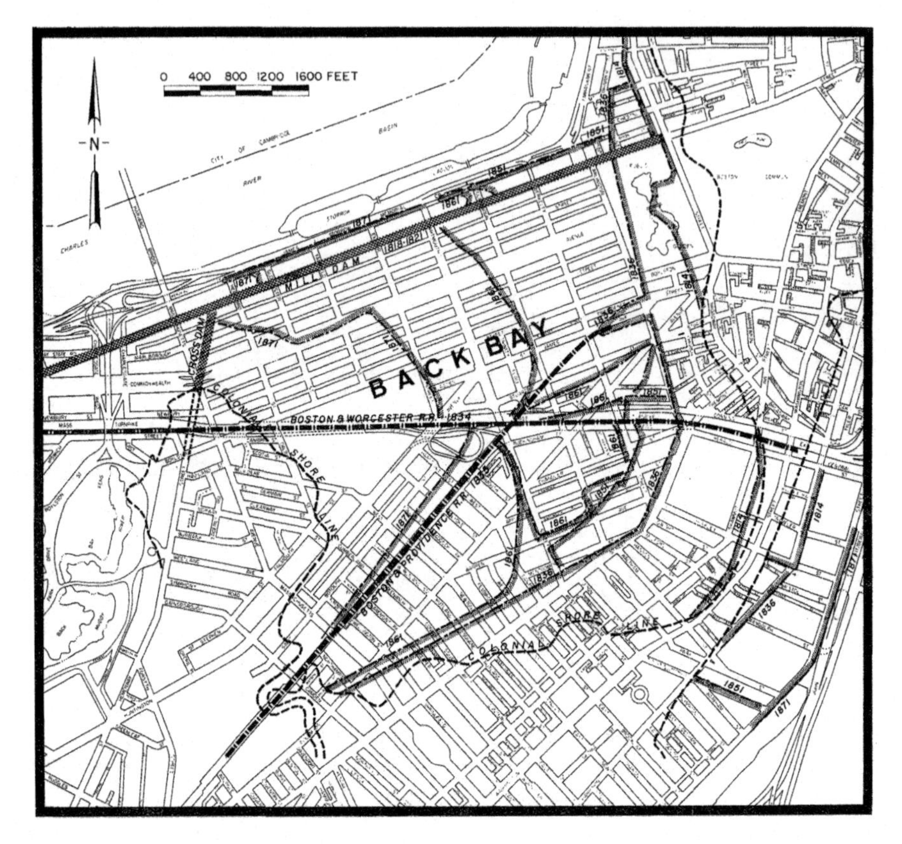

Figure 3.3 Sequence of filling in Back Bay, Boston, Massachusetts.

If this project were proposed today, it would have never been built because of environmental restrictions on filling of tidal lands. Also, the public process would have resulted in a planning stage that would have taken years, not months to accomplish.

Tunneling then and now

Our understanding of geotechnical principles related to tunnel construction grew rapidly in the late 19th and early 20th centuries as tunnels were built in major cities such as Boston, New York, Detroit, and Chicago for subways, railroads, highways, and water and sewage lines. Successful completion of these projects required an expanded understanding of the principles of ground behavior varying from very soft river and marine sediments to bedrock.

For example, the St. Clair River Tunnel between Port Huron, Michigan and Sarnia, Ontario was constructed in 1888 for the Grand Trunk Railroad through very soft soils. The tunnel was built using a compressed air shield that is illustrated in Figure 3.6.[4]

Figure 3.4 John Souther's steam shovel in Needham, Massachusetts loading gravel for filling of Back Bay.

Figure 3.5 Charles River Basin in 2000.

The shield provided immediate support of the soil at the tunneling face for protection of workers and also provided a space in which ground supports could be erected. The compressed air helped maintain a stable working face and was used to minimize groundwater inflow. Similar equipment was being used at the time in other major cities such as New York and London.

In the late 19th century in Detroit, the desire to create railroad and highway links to Windsor, Ontario under the Detroit River led engineers to combine innovative tunneling techniques to effectively accommodate geotechnical conditions along the tunnel shore approaches and under the river. The Detroit-Windsor Rail Tunnel illustrated in Figures 3.7[5] and 3.8[5] was the first sunken tube tunnel constructed in North America.

The sunken tube tunnel construction method in common use today, was a totally new and innovative approach in the early 1900's. Water depths and subsurface conditions in the middle of the river dictated the adoption of this method of construction. The maximum river depth was approximately 25 feet with 45 feet of soil overlying limestone bedrock. The soil profile consisted of about 15 feet of very soft bottom sediment overlying 20 feet or so of medium stiff lake clays that in turn overlay a very dense clay-like glacial till. It has long been understood that tunnel construction is much more effective and less costly when constructed totally in soil or totally in rock. To build tunnels at the soil rock interface is difficult and costly because soil–tunnel construction methods are different from rock–tunnel construction methods. To build a tunnel partially in soil and partially in rock is possible, but typically very difficult because two different approaches to construction must be blended, leading to inherent inefficiencies and

Figure 3.6 Compressed air shield – St. Clair Tunnel (1888).

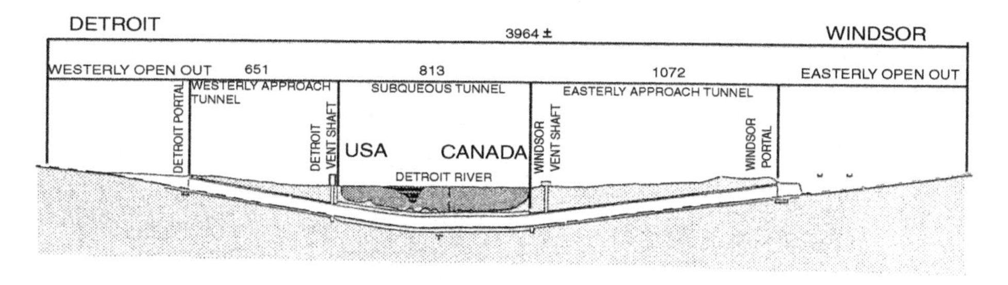

Figure 3.7 Profile of Detroit Windsor Rail Tunnel (1909).

increased cost. However, when done, this is known as "mixed-face" tunneling. Conditions under the Detroit River were such that there was insufficient soil cover to permit shield tunneling with air. There were only 15 feet of soft soil cover that would probably have been too little to contain the air pressure

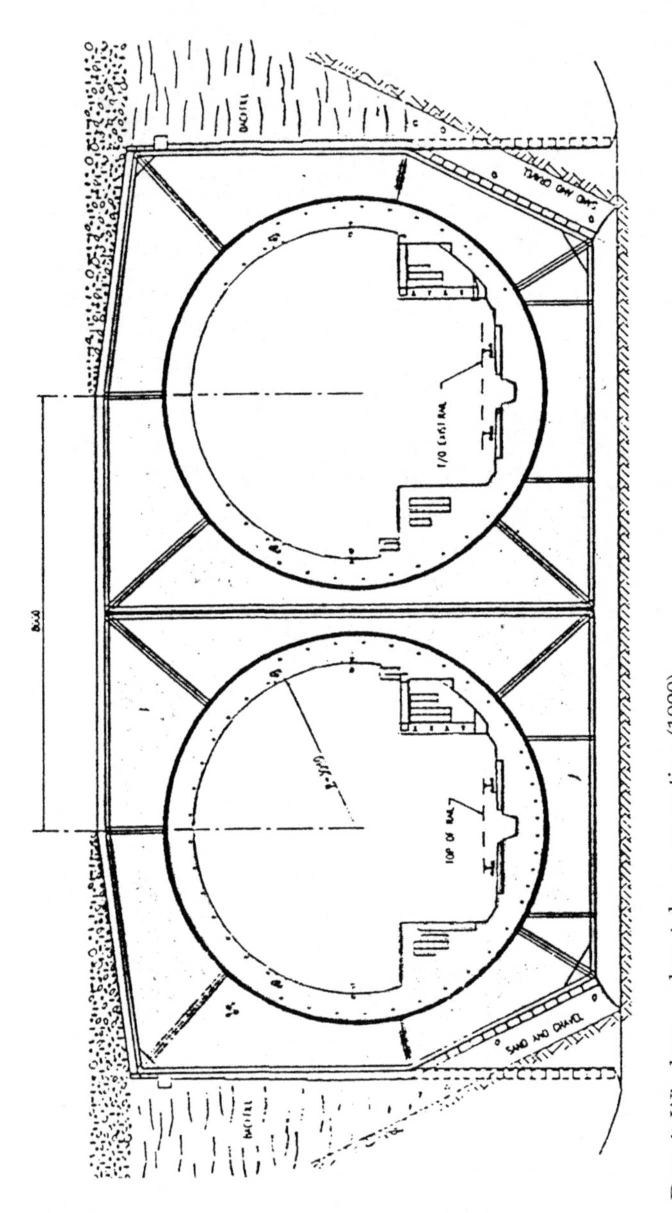

Figure 3.8 Detroit-Windsor sunken tube cross-section (1909).

thereby resulting in a "blow" into the river with catastrophic results. To build the tunnel entirely in rock would have required a very deep alignment with very long approach tunnels on either shore. The project engineers came up with the idea of excavating a trench in the river bottom, floating a prefabricated steel tunnel section over the trench, and sinking it in place onto the trench bottom. This was done in 50 feet long segments that were joined together forming watertight seals, and backfilled with soil and concrete. The concrete backfill and soil cover were designed to provide sufficient weight so that the tunnel sections would not float when the completed tunnel was dewatered to permit construction of the final lining and tracks.

The approach tunnels under either shore were constructed as shield driven tunnels to a point at which the amount of cover was insufficient for tunneling and a concrete box was built and buried by what is referred to as the "cut-and-cover" tunnel construction method. The final portions of the approach to join the tracks at existing ground surface were completed in open cuts both with and without soil retaining walls.

If this tunnel were to be built today, it would probably not be done by the sunken-tube method because of environmental concerns with river water quality issues. The bottom sediments are contaminated as they probably were in the early 1900's, but at that time there was no concern for degradation of water quality due to soil dredging with a resultant muddy plume traveling down river, nor was there concern for the bottom habitat of fresh water organisms. Today, there are numerous regulations that either would prevent such construction out-right or as a minimum require so many environmental controls that the project would be economically infeasible.

The economic needs for increased rail traffic between Michigan and Ontario dictated that another tunnel be built under the river. Therefore, in 1996 a similar railroad tunnel was completed under the river at a location about 60 miles north of this project through very similar geotechnical conditions. However, this project was built as a bored tunnel entirely in soil without compressed air. Sunken tube construction was considered, but largely rejected on environmental grounds for reasons discussed above. It was built with an Earth Pressure Balance (EPB) shield in combination with a pre-cast concrete segmental liner system installed as excavation advanced. A typical EPB machine is illustrated in Figure 3.9.[6]

EPB shields allow tunnel construction in soft soils without the need for compressed air. The shield is a closed face system by which soil stability is maintained behind a bulkhead. Workers operate in atmospheric conditions and no contaminated surface soils are excavated. This eliminates many environmental restrictions on the project. This technology, initially developed in England in the 1960's has been refined over the ensuing decades and applied on a large scale mostly in Japan. Today, the technology is widely used throughout the world. The machine used to construct the Sarnia Tunnel was built in Canada.

Geotechnical issues have largely driven all of the above described tunnel construction projects. Environmental considerations were largely not a

Figure 3.9 Earth Pressure Balance Shield cross-section.

concern until the 1970's when a dramatic increase in regulations resulted in such issues often controlling the design and construction process and, in some instances, making projects infeasible.

Buildings and infrastructure; then and now

Cities are population centers comprised of buildings of all types and all interconnected by a maze of both surface and subsurface infrastructure such as roads, bridges, water lines, sewers, gas lines, plus buried power and telecommunication cables. Design and construction of all of these facilities requires examination of geotechnical and environmental issues because they all are either supported on, or are constructed within, the ground.

Excavation challenges

The effective use of underground space is essential to the ultimate viability of cities. This is especially true in the biggest cities, i.e. "Mega-Cities" such as Mexico City, Bangkok, and Los Angeles. Such cities would be almost uninhabitable without significant underground transportation, utility, and storage spaces. Just about all urban buildings include basement space that houses mechanical systems, storage, and frequently parking. To build infrastructure or to create basement space requires excavations adjacent to buildings, streets and utility lines. These building excavations are typically 10 to 25 feet deep (Figure 3.10) which is an adequate depth for two basement levels, but depths of 40 – 60 feet are not uncommon. While not typically associated with new buildings, excavations over 100 feet deep are being created typically for transportation facilities such as Boston's Central Artery/Tunnel Project or for subway stations as illustrated in Figure 3.11.

Figure 3.10 Overview of braced excavation, Boston, Massachusetts.

Excavations in soil and/or rock will always result in some horizontally inward and vertically downward movement of the ground adjacent to the excavation for a distance approximately equal to the depth of excavation as illustrated in Figures 3.12[7] and 3.13.[8]

In rock and stiff soils, these movements may be very small and not extend all that far from the limit of excavation. However, in loose and soft soils, the movements can be substantial, i.e. several feet, and in extreme cases the sides of an excavation may collapse because the excavation support system is not strong enough to withstand the excavation induced lateral pressures.

Excavations are an essential element of urban construction projects. They are typically made adjacent to existing buildings, streets, and/or buried utilities which must be protected. Preventing damage to these facilities typically requires lateral support of the sides of excavations and underpinning of adjacent buildings to minimize horizontal and vertical movement to acceptable levels. A typical tie-back support system is illustrated schematically in Figure 3.14.[9]

Lateral support can vary from timber braces to steel soldier piles with timber lagging, to interlocking steel sheet piles or cast-in-place reinforced

Figure 3.11 A deep internally braced excavation in Boston, MA.

concrete slurry walls. Selection of appropriate methods is a function of excavation depth, soil, rock, and groundwater conditions and cost. The applicability of various methods is summarized in Table 3.1.

An excellent example of extensive lateral support was the seven–level underground garage at Post Office Square in Boston constructed in 1989. This project covered an entire city block and required excavation to an

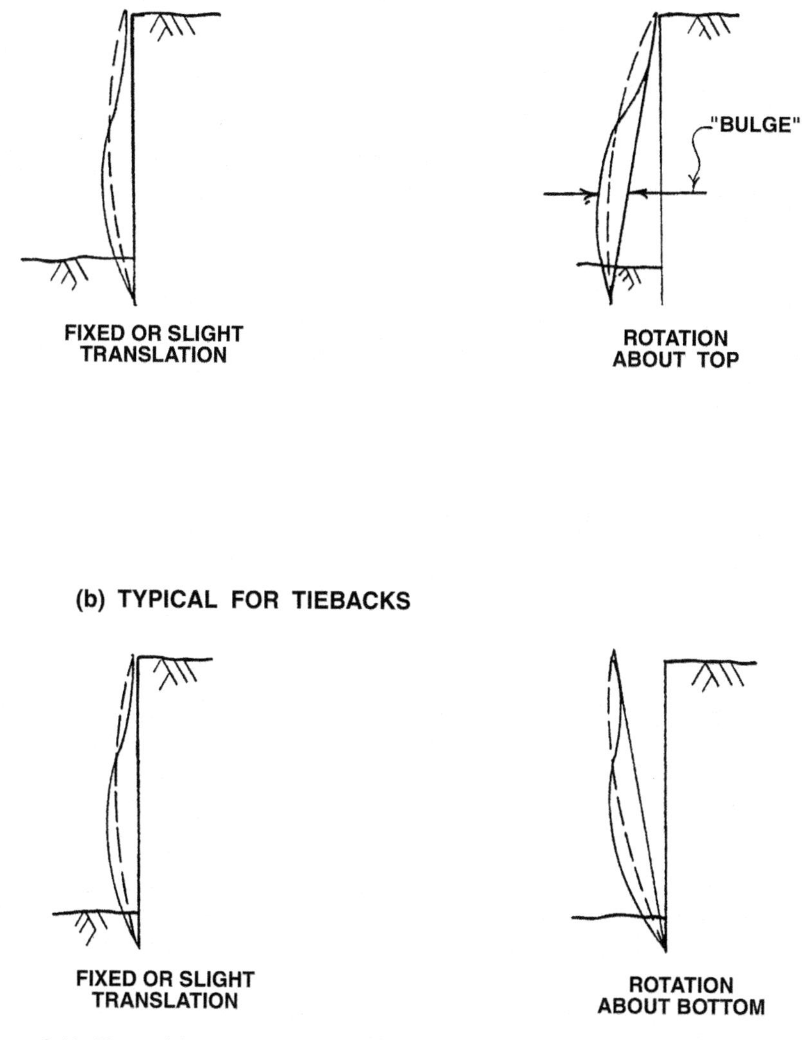

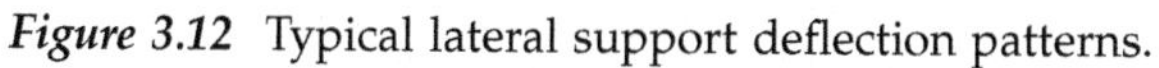

Figure 3.12 Typical lateral support deflection patterns.

approximate average depth of 85 feet. More than 60,000 square feet of cast-in-place reinforced concrete slurry walls were employed as the primary support of this excavation method. These slurry walls were some of the deepest walls constructed by the slurry wall construction technique in North America at the time. The slurry wall method of construction is one that permits construction of reinforced concrete walls in place prior to general excavation. Figure 3.15[10] is a simplified cross-section through the project.

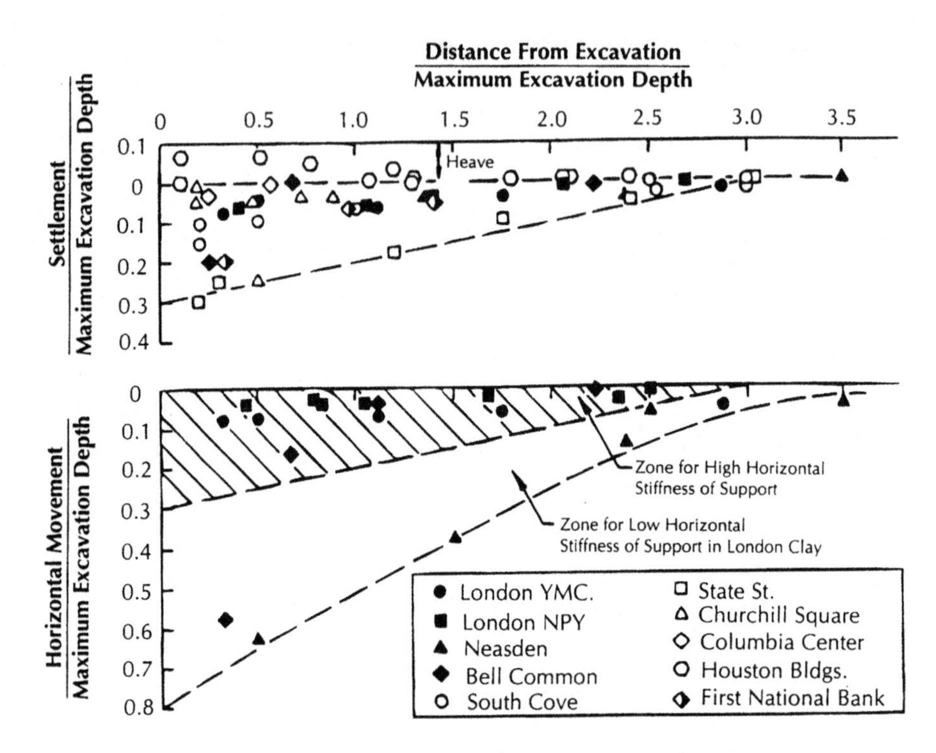

Figure 3.13 Ground deformation patterns adjacent to excavations.

The slurry wall method was developed in Europe and has been used extensively in North America since about 1964. The method facilitates construction of a reinforced concrete wall in the ground prior to general excavation. Through the use of this method, it is often possible to eliminate underpinning of adjacent buildings and dewatering because the relatively stiff wall, if properly braced, prevents lateral ground movement and the continuous wall is an effective groundwater cutoff if founded in impermeable soil or rock. An automated slurry wall excavator is illustrated in Figure 3.16. A trench, typically 3 to 4 feet wide, is excavated in soil and kept filled with a slurry consisting of either bentonite and water or a polymer and water to maintain the excavation stability. The trench excavation is done in alternating sections.

Once the trench is excavated, reinforcing steel consisting of steel beams, reinforced steel cages or both are placed and secured at the proper position in the slurry-filled trench. The slurry is displaced as concrete is placed in the trench starting at the bottom using a tremie pipe. Once a panel is completed in this way and the concrete has attained the required strength, alternate panels are excavated to create a continuous wall in place. General excavation can now begin and the wall constructed in place

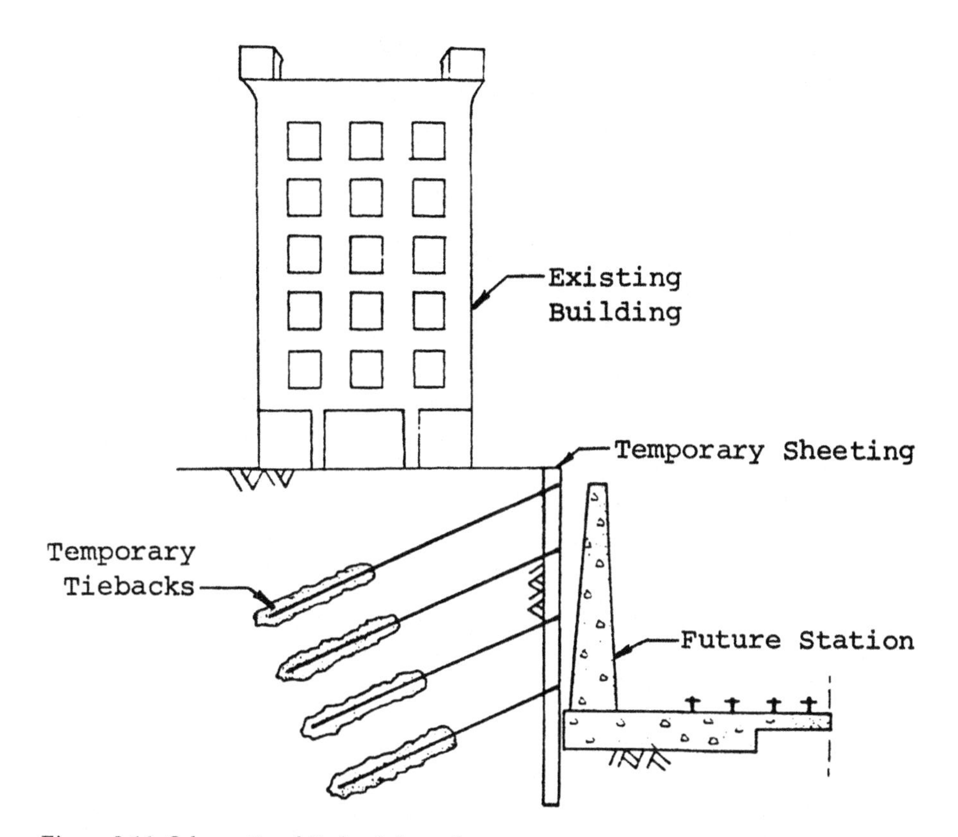

Figure 3.14 Schematic of tie-back lateral support system.

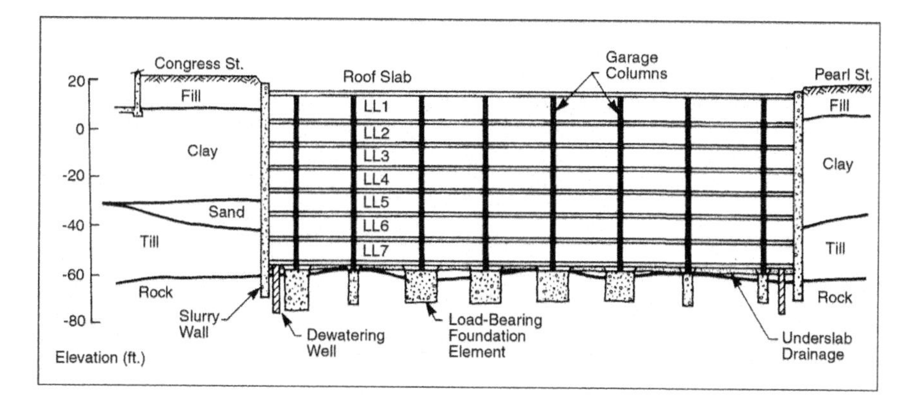

Figure 3.15 Cross-section through Post Office Square garage excavation in Boston, MA.

Table 3.1 Summary of Excavation Support Methods

Type of Support	Applicability Criteria				Overview Comments
	Soil/Rock Conditions	Groundwater Conditions	Excavation Depths	Underpinning Required?	
Timber shoring	Granular and stiff cohesive soils.	At or below the bottom of excavation. Dewatering of granular soils generally required if groundwater is above bottom of excavation.	5 feet to 15 feet	Yes, if adjacent buildings and utilities are close to excavation.	Simple excavations for small buildings and utility trenches.
Soldier Piles and Lagging	Granular or stiff cohesive soils.	See remarks above for timber shoring.	40 feet to 50 feet	Yes, if adjacent buildings and utilities are close to excavation, but underpinning can be minimized and possibly eliminated if the structural system is very stiff.	
Steel Sheet Piling	All soil types except very dense soils or strata with frequent boulders.	Dewatering can be eliminated as long as pile interlocks, i.e., vertical joints, do not leak excessively.	10 feet to 70 feet	See comment above for soldier piles and lagging.	
Slurry Walls	All soil types. Walls can be installed in rock formations.	Generally not required except for protection against bottom heave due to underlying pervious strata over impervious strata in bottom of excavation.	40 feet to >100 feet	Generally not required as long as wall is sufficiently stiff.	Slurry walls are frequently the permanent foundation wall. No cast-in-place wall is required after excavation. An architectural treatment may be required to cover the rough surface of the slurry wall depending on the application.

Figure 3.16 Slurry wall schematic.

can be used as the permanent foundation wall thereby eliminating construction of a double-formed cast-in-place wall inside of a temporary support of excavation system.

Underpinning challenges

In some instances lateral support is insufficient to protect adjacent buildings, streets or utilities from damage due to ground movement. Where this occurs, the affected facilities must be underpinned. Underpinning is typically done in one of two fundamental ways. The first and most common approach is to provide a new structural support that transfers loads down to a level where movement is not occurring or is at tolerable magnitudes. This has been done in various ways for hundreds of years . For many years, the most common way that structural support was provided prior to excavation was by "pit-underpinning" or by jacked short piles with or without "needle beams." "Pit-underpinning" is a labor-intensive tedious process that involves hand excavation of small pits below structure support points down to the desired bearing level. The column loads are temporarily transferred by means of "needle-beams" to other locations as the pits are made. The pit walls are typically supported by steel timber shoring. After the desired depth is reached, the pits are backfilled with concrete up to the underside of the structure foundation. The process is repeated at each structural support point (Figure 3.17).[7]

In recent years, underpinning is more commonly done using micropiles or by ground improvement through grout injection or freezing. Micropiles

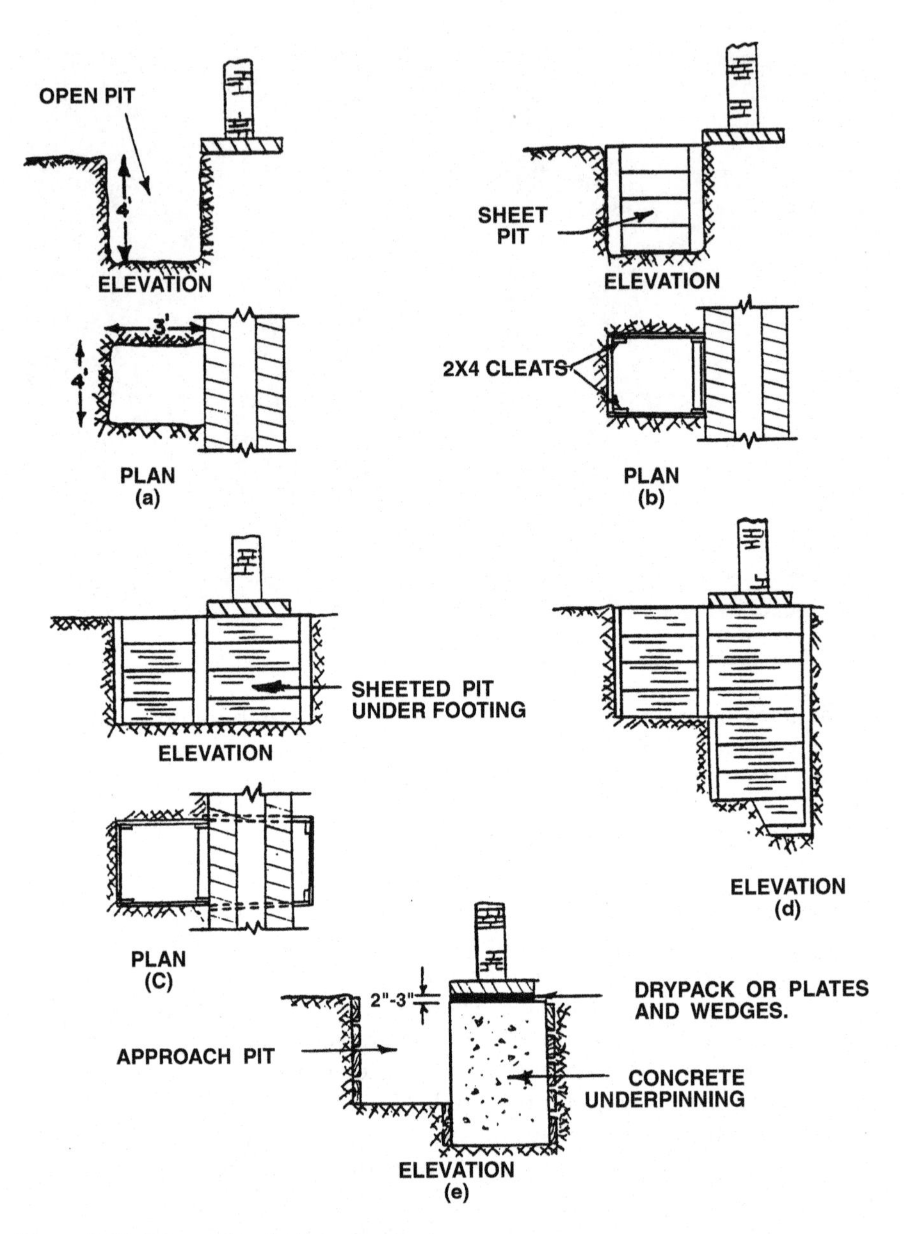

Figure 3.17 Schematic of pit underpinning.

are, generically, small diameter, bored, grouted-in-place piles incorporating steel reinforcement.[11] They are installed through the foundation to be underpinned to the desired bearing level. The application of this method requires sufficient headroom to accommodate the drilling equipment. Special drilling equipment has been developed by several manufacturers that can operate in as little as 5 feet to 6 feet of headroom.

The second fundamental approach is to improve the ground properties, i.e., stiffen and/or strengthen the ground, to prevent or minimize movement. In either case, the underpinning treatment may or may not be carried below the bottom of the excavation depending on the distance of the facility from the excavation and the subsurface conditions. This can be done in a number of ways, with two common approaches being cement grouting, and ground freezing. A schematic illustration of grouting for underpinning purposes is illustrated in Figure 3.18.[12]

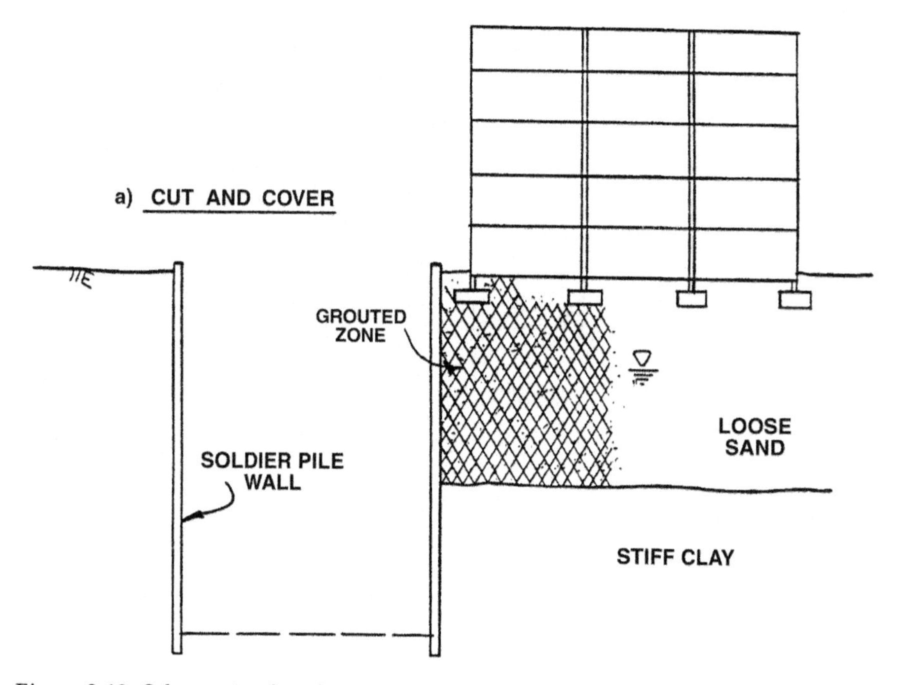

Figure 3.18 Schematic of underpinning by ground improvement.

These approaches are typically applicable in situations where the loads being supported are not great, such as for one to two story buildings. Underpinning of larger structures typically requires a more positive structural method such as micropiles as described above. Cement grouting can be effective because it improves the engineering properties of soil by filling the void spaces with cement slurry that sets to create a stable soil/cement mass.

Freezing of soil in place is a less common approach, but it can be effective for temporary situations such as fine-grained soils that cannot be grouted. The soil is frozen by installing brine–filled refrigerant pipes in drill holes throughout the soil mass to be frozen. The brine system freezes pore water in the soil mass thereby improving the soil's properties in a manner similar to the grouting process. Soil freezing must be maintained for as long as the underpinning is required. Care must be taken to monitor ground movement during freezing operations because the freezing of soil often results in soil volume change that can cause heave during freezing and settlement as the frozen soil thaws. Underpinning by soil freezing is usually only done in special situations requiring unusual approaches to project design and construction. One such case is the proposed construction of transit tunnels under the Russia Wharf Building in Boston. In this situation, the soil will be frozen to permit the support of existing timber piles as they are cut off to permit tunnel construction directly below portions of the building. Figure 3.19[13] illustrates a typical; section for this project.

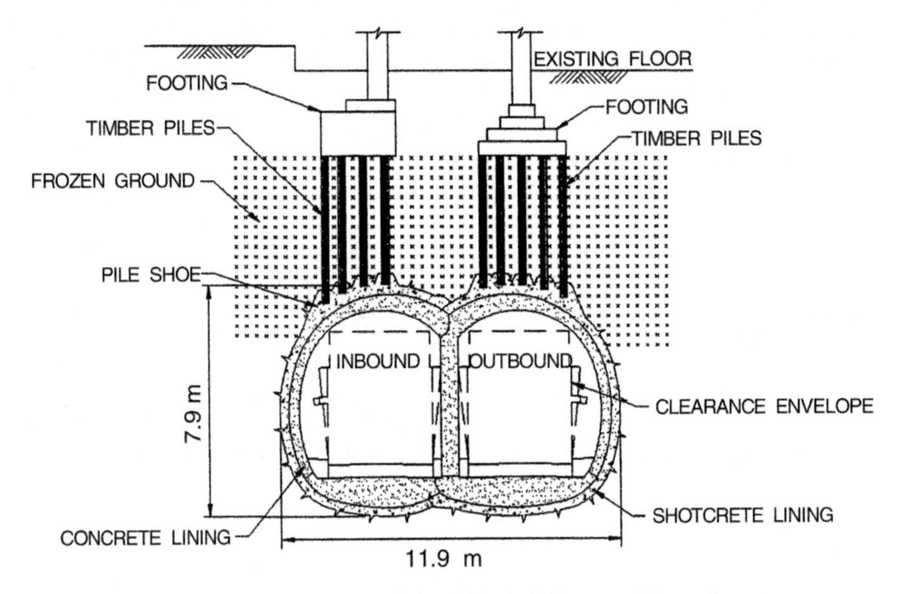

Figure 3.19 Cross-section of Russia Wharf Tunnel, Boston, Massachusetts.

The Russia Wharf project design is an illustration of the fact that almost any subsurface technical challenge can be overcome through ingenous technical approaches, however, such approaches can be very expensive.

The interrelationship of geotechnical and environmental issues

As discussed above, geotechnical and environmental issues are almost always interrelated. It is generally not possible to work in or on the ground without consideration for environmental regulations whether in a city or not. An example of this interrelationship is the trend to provide more expensive deep foundations; typically driven piles, to transfer building loads through contaminated soils to competent soil at modest depths in lieu of less expensive spread foundations constructed in an open excavation. The driven pile option nearly eliminates the need to excavate and therefore dispose of contaminated soil, usually urban fill. Depending on the level of contamination, regulations will permit leaving moderate to low levels of contamination under buildings as long as the contamination is contained. The disposal of contaminated soil, even soil with relatively low levels of contamination, has to be done as specified by complex regulations that can lead to unacceptable project delays. Therefore, many buildings are now supported on driven piles or some other form of non-excavated deep foundations rather than on spread footings as would have been done prior to current environmental regulations.

Another good example of this interrelationship is groundwater control in tunnel and foundation excavations. Assuming the groundwater levels must not be maintained for other reasons such as deterioration of wood piles or area settlement due to compressible soils, it is usually cheaper to pump and dispose of groundwater than it is to exclude it from the excavation. However, if the groundwater is contaminated, even at relatively low levels, it is then usually cheaper, faster and simpler to design a system whereby groundwater is excluded from the excavation. This can be accomplished by slurry walls carried to impervious soil or rock strata below the excavation or by tunneling methods such as earth-pressure balance or slurry shield tunnel boring machines. In summary, the guiding design and construction principal is to avoid handling contaminated soil, rock or groundwater whenever possible.

Summary

This has been only a brief overview of some of the geotechnical and environmental challenges that must be overcome when construction in a city is planned. Over the past 30 to 40 years, the process has become much more complex as a result of environmental regulation and control. These processes frequently add years to project schedules. One of the very best examples is the design and construction of the Charles River Dam in Boston for which complete engineering and environmental studies were completed in nine months! If this project were proposed in current times the environmental and design schedule would be measured in an indeterminate number of years. The wisdom and benefit to current aggressive environmental regulation is a subject of considerable on-going debate as society struggles with the competing challenges of creating a workable urban environment while maintaining reasonable environmental conditions.

At one time in the past, geotechnical and environmental issues were considered independently, but this is no longer possible. They are closely interrelated and must be considered together. It is no longer acceptable for engineers and environmentalists to be uninformed of each other's disciplines. The process is increasingly complex, both on a technical level and on a regulatory/political level.

References

1. Figure 3.2 courtesy of the Boston Athenæum, Boston, MA.
2. Aldrich, H.P., Back Bay Boston, Part 1, *J. Boston Soc. Civil Eng.*, Boston, MA, January 1970.
3. Figure 3.4 courtesy of the Boston Athenæum, Boston, MA.
4. Mathews, A.A., *Tunnel Shields for Subaqueous Works*, A.A. Mathews, Inc., Santa Clara, CA., November 1968.
5. Garrod, B.L., et al., The Detroit River Tunnel Remedial Grouting Program, 1993 Annual Publication, *The Tunnelling Association of Canada*, 1993.
6. Figure 3.11 courtesy of Lovat, Inc., Etobicoke, Ontario, CA.
7. Goldberg, D.T., et al., Lateral Support and Underpinning, *Federal Highway Administration Research Report FHWA-RD-128*, Washington, D.C., April 1976.
8. Peck, R.P., "Deep Excavations and Tunneling in Soft Ground", *Proc. 7th Intl. Conf. Soil Mechan. Foundat. Eng.*, Mexico City, State of the Art Volume, 1969.
9. Weatherby, E.E. and Tiebacks, E., *Federal Highway Administration Research Report FHWA/RD-047*, Washington, D.C., July 1982.
10. Erickson, C.M., et al., Predictions and Observations of Groundwater Conditions During a Deep Excavation in Boston, *J. Boston Soc. Civil Eng.*, Boston, MA, Fall/Winter 1993.
11. Bruce, D.A., et al., High Capacity Micropiles — Basic Principles and Case Histories, *Proc. Third National Conf. Geo-Institute Amer. Soc. Civil Eng.*, Urbana-Champaign, IL, June 1999.
12. Guertin, J.D. et al. (1982), Groundwater Control in Tunneling, *Federal Highway Administration Research Report FHWA-RD-81/075*, Washington, D.C., April 1982.
13. Gall, V., et al., Frozen Ground for Building Support — Implementation of Innovative Engineering Concepts for Tunneling at Russia Wharf, *Proc. N. Amer. Tunnel. 2000*, Boston, MA, June 2000.

chapter four

The contest with groundwater for underground space

J. Patrick Powers

Contents

Groundwater and construction 49
Controlling groundwater while we dig and build 50
The methods of groundwater control 51
Structure design below the water table 53
A plethora of choices 53
Cost of controlling groundwater 54
To pump or not to pump 54
Harming the flora 56
Wetlands 56
Summary 57

Groundwater and construction

At some depth below the land surface, our excavations encounter the water table, the flat or usually sloping surface beneath which all the pores and fissures of the earth are saturated. Groundwater has been a precious resource to mankind since before the dawn of history. Many of our great cities were sited where they are by our ancestors who found fresh water there; in lakes or rivers, or flowing up from the ground in springs. In modern times, the groundwater, once a boon, becomes a problem as our development of underground space takes us deeper below the water table. The problem has two prongs: we must control the groundwater while we excavate and build the

0-8493-7486-3/01/$0.00+$.50

structure; and we must design the structure so it can resist the forces of pressure and buoyancy that groundwater will exert upon it.

Controlling groundwater while we dig and build

There are two basic methods for controlling groundwater: we can pump the water, which lowers the water table so the work can proceed; or, we can prevent the water from entering the excavation by one of the ingenious cutoffs builders and engineers have developed through many decades. There are scores of variations, but all the control schemes involve one of the two basic methods, or a combination of both.

Some of the schemes for controlling water are old. The biblical well of Jacob must have involved some pumping during its excavation, perhaps with a leather bucket and rope. In the 1700's, the English coal mines were pumped first by treadmills, then primitive steam engines, as miners followed the seams deeper into the earth. Cutoff methods were being used in the nineteenth century, for mining, and occasionally for construction excavations. Today the traditional methods have been dramatically improved, and new variations continue to emerge. The technology for applying the various methods is well known; it was developed mainly by trial and error, combined with study and analysis of the many failures that happened along the way. Groundwater has traditionally been surrounded by mystery. To this day, dowsers walk about with forked sticks, believing they can find it. As late as the middle of the twentieth century, some United States courts refused to rule in groundwater disputes, holding that movement of water within the earth was unknowable.

But determined men spurred by the economics of groundwater have dispelled much of the mystery. Men seeking groundwater for thirsty citizens have developed analytic methods that when skillfully applied can select favorable sites for water wells, predict their capacity, and how long they will last. This technology has been adapted by construction engineers, who added some variations to suit their need to control groundwater. Today it is possible to predict with reasonable reliability how the groundwater will act, and which among the available methods are suitable for a given site. Not all the mystery is gone, however. Whenever you dig below the land surface, there is some degree of uncertainty. The successful practitioner in groundwater control bases his design on soil borings and appropriate field and laboratory tests; he understands the characteristics of the various methods at his disposal, and he keeps his design flexible, so it can be modified if conditions encountered are different than expected.

In this chapter, I shall describe the methods of groundwater control, and list some of the advantages and disadvantages of each. I will also discuss the impact groundwater control can have on the cost and schedule of a project, on the regulations that govern applications, and the potential damage that can sometimes occur to neighbors, and the environment. Such damage can be avoided if its potential is evaluated reliably in advance, and appropriate preventive measures are undertaken.

Also we will look at the impact groundwater has on the design of a structure. If the water table has been lowered by pumping during excavation and construction, when pumping ceases the water table will gradually rise to its original level. We will discuss the means that have been used to build structures capable of withstanding the resulting stress.

The methods of groundwater control

The oldest and most direct method of controlling groundwater is to let it flow into the excavation as it proceeds, collect the water in ditches, and pump it away. The process is called open pumping. In dense, stable soils and fissured rock, the method can be effective. But in sensitive uniform fine sands, the results can be catastrophic. Perhaps you have watched a toddler on the beach with his pail and shovel digging near the water's edge. If he reaches the water table he is puzzled. Water flowing into his excavation carries sand with it, and no matter how fast he digs he cannot get deeper. We may smile patronizingly at the toddler, who doesn't understand about water tables. But I have seen a contractor with a seven cubic yard dragline who was making no more progress in running sand than if he had a pail and shovel.

When open pumping is unsatisfactory, we can surround the proposed excavation with pumping devices such as water wells or wellpoints. Pumping these devices can lower the water table before we dig, in the process called predrainage. Under favorable conditions, we have seen excavations carried forty feet and more below the original water table, with no water visible, except at the discharge pipe from the pumping system.

If, on a given project, cutoff is preferred to pumping, there are many methods available. Before the turn of the century, Wakefield Sheeting, a composite board with a tongue and groove shape, gave reasonable performance once in place, but had difficulty surviving installation except in very soft ground. It is rarely used today.

Steel sheet piling has been around since the nineteenth century, and continues in widespread use today. The steel sheets or bars are available in various strengths. There is a ball-edge along one side and a socket edge along the other so when driven into position with a pile driver, the sheets interlock. The sheeting is not watertight, but if none of the sheets have jumped out of interlock while being driven, the leakage is usually moderate.

Rebuilding the devastated infrastructure in Italy after World War II was hampered by the shortage of steel for sheet piling. Ingenious contractors came up with a way to build a vertical concrete barrier underground, called a slurry wall or diaphragm wall. Short trenches, panels as they are called, were excavated while being kept filled with a slurry of bentonite clay. That kept the sides of the panel from collapsing, for reasons that are still not understood fully. After the panel is excavated, it is filled with concrete from the bottom up. Different methods were developed for constructing joints between panels. The method was first applied in the United States in the 1960s for New York's mammoth World Trade Center, and its use has since become widespread.

Circular cast in place concrete piles, drilled in by the slurry method have been used for cutoff. Piles that touch are called tangent piles; piles that intersect are called secant piles. The latter are better able to create a watertight wall.

In the process called jet grouting, a vertical pipe with a horizontal nozzle injects a cement slurry into the soil, while the pipe is rotated and raised sequentially. The result is a cylinder of soil mixed with cement. In a variation, the original soil is removed and replaced with a sand/cement mixture. A line of such cylinders are spaced so they intersect each other and form the cutoff.

In recent years, the soil mixed wall has appeared. A gang of augers held together, drills into the soil, injecting cement into the augured soil. Successive panels form the wall. Steel reinforcement with H-beams can be provided.

All the cutoff methods just described can be braced suitably or tied back, and used as ground support, maintaining the vertical sides of an excavation, besides cutting off water flow.

Ground freezing, in which chilled brine is circulated through steel pipes in the ground until the pore water in the soil freezes, has been used for deep coal mine shafts in Britain for well over a hundred years. Today it is widely used in construction and mining in North America. Ground freezing can support the earth and provide cutoff. The process is most efficient with circular section excavations such as tunnels and shafts.

Permeation grouting, in which a chemical liquid of low viscosity is used to permeate the soil pores, and then is hardened or gelled by a chemical process, has been in use for many decades. Fully effective cutoff is difficult to accomplish; if the grout is pumped too rapidly at too much pressure, it will fracture the ground instead of permeating the pores as desired. Modern instruments to observe pressure, flow and other factors, and modern computers to interpret the observations, have enabled skilled practitioners to improve quality control and achieve effective cutoffs.

All the previous cutoffs are oriented vertically. In typical practice, the cutoff is extended to seal in some impermeable layer such as clay or rock. If no such layer exists at a reasonable depth, a horizontal cutoff may be advisable. The tremie concrete seal has been in use so long its origins are forgotten. This excavation is surrounded by a vertical cutoff, typically steel sheet piling or a diaphragm wall, and the digging is done under water with a clamshell bucket. A concrete slab is poured under water through a tremie pipe. The slab must be thick enough to resist the buoyant forces. When the concrete sets up the water is pumped out. The recent improvements in permeation grouting described earlier have made it possible to build a horizontal grout blanket. Jet grouting also has been used for making horizontal cutoffs.

Compressed air, while it is not really a cutoff, serves the same purpose by excluding groundwater from tunnel excavations. Using a system of compressors and locks, air pressure inside the tunnel is maintained at a high level to prevent water from entering in unmanageable quantity. The same process can be used for sinking vertical shafts; in the 1870's, Washington Roebling used this to build the foundations for the towers of the Brooklyn Bridge.

Tunnel boring machines incorporating earth pressure balance features, regulate the entry of soil and water through the face in coordination with the rate of tunnel advance, and full dewatering is not required. Sometimes partial dewatering is used to reduce the stress on the machine. In a rare application, several years ago water wells were put in the bed of the North Sea, to relieve the stress on an earth pressure machine.

Structure design below the water table

Structures below the water table need special design features. They must be watertight, or the underground space will be wet and of limited usefulness. The pressure of the water seeking entry increases with depth. For deep structures, elaborate systems of membrane waterproofing and seals at construction joints have been developed. The foundation slab and walls must have strength to resist the hydrostatic pressure exerted by the water.

The buoyant force created by the water must be resisted. With five levels of underground parking for example, even a forty story building may lack the weight to prevent flotation. A permanent pressure relief system is often used, particularly in dry docks. The base slab thickness may be increased but since one foot of concrete below the water table resists only one-and-a-half feet of buoyancy in water, the cost of the concrete and the extra excavation escalates. Soil and rock anchors sealed in cement grout are often used to help hold the structure down. Sometimes designers rely on the buoyancy to carry part of the foundation load of a heavy structure. In such cases, the recovery of the water table must be carefully monitored and controlled. Pumping must continue until the structure is heavy enough to resist flotation, and then the water table must recover at a preestablished rate. If natural recovery is too slow, artificial recharge may be advisable to avoid delay.

A plethora of choices

We can see that an engineer or contractor confronting a project that goes below the water table has a great many options from which they might choose. The necessary first step is to investigate the underground at the site, taking soil borings, performing field and laboratory tests, and researching previous experience in the area. You should study any nearby structures, including underground utilities, to see if they may affect your work, or be affected by it. On land that once had been an industrial site, you should be alert for evidence of contaminated groundwater. You need to be familiar with local, state and federal regulations which might apply to groundwater. With all that in hand, you must familiarize yourself with the many methods of groundwater control suitable for the site, before you choose among them.

Some examples: steel sheet piling is unsuitable in ground containing many boulders. Boulders cause hard driving, and may result in split sheets, jumped interlocks and other problems. A slurry wall can get through bouldery ground, by breaking them (with a heavy chisel) into pieces small enough to be fished out of the panel. But that is time consuming and expensive. Often the optimum solution is to support the ground with soldier piles and timber lagging. The lagging does not cut off the water, so the site must be dewatered.

Cost of controlling groundwater

Whenever you hear stated at a meeting "money is no object," you can conclude the speaker lacks understanding. Cost is always a concern, in selecting the method to build a project, or whether to build it at all. In one of our major cities, a deep underground parking garage to be built under difficult soil and water conditions was under study. "We must have the parking!" the architect/planner pronounced. A team of engineers analyzed the problems and concluded the garage could be built using techniques within the existing state-of-the-art. They also prepared an estimate. The owner was advised he could have his garage, if he could afford sixty thousand 1965 dollars per parking space. He could not justify such a cost, and considered himself fortunate to have been apprised of the cost in advance. If the difficulty of an underground project is misjudged, and the cost underestimated, adjusting after construction begins can be enormously expensive. Cost estimation is a vital part of the design process, if delays, cost overruns, and the problems that lead us into litigation are to be avoided.

To pump or not to pump

Lowering the water table on the majority of projects is the least–cost approach to controlling groundwater. But under certain specific conditions, pumping can have undesirable side effects. In my fifty-odd years of experience in groundwater control, the frequency of encountering such specific conditions has been quite small. In cases where one or more of the conditions has been encountered, means have been found to mitigate the side effects at significantly less cost than building the project without dewatering. But sometimes under severe conditions, it may be the better course to build within cutoffs, so pumping is limited, or eliminated.

Some of the conditions that may cause side effects are:

Contaminated groundwater. A great many organic and inorganic chemicals have been spilled in our urban areas, and many have leached into the groundwater. Contaminated discharge from a dewatering system must be treated before it can be released into the surface water environment. The treatment cost varies, depending on the contaminant. But the combined cost of pumping and treating is often less than the cost of cutoffs. The pump and treat method has the very significant advantage of cleaning up the contaminant plumes.

Depleted groundwater supply. If the aquifer in which the groundwater must be controlled is being used for the water supply, dewatering may cause at least a temporary reduction in water well capacity, and if near the seacoast may aggravate salt water intrusion. The potential problem should be studied by a qualified water supply hydrologist. The dewatering discharge on some projects has been discharged back into the ground. For others, temporary alternate supplies have been provided to the users.

Ground Settlement. Dewatering places a modest load on the subsoil, because it reduces buoyancy. Most soils at depth have the strength to absorb this load without consolidation that might damage existing structures. My experience has been that in many cases, concern over consolidation is unwarranted. In most urban areas, the subsoil has already experienced loads greater than the increment from dewatering. But if there are weak, compressible soils near the dewatering operation, undesirable consolidation has occurred. The problem can be mitigated by partial cutoffs, artificial recharge, or in some cases partial penetration of the dewatering system into the aquifer. The technology exists for analyzing the potential problem by field and laboratory tests. If the problem is confirmed and the mitigating methods are unsuitable, it may be advisable to construct the project without dewatering. In a well documented project thirty years ago, a depressed section of an interstate highway project was to be built through downtown Sacramento, California. A major pumping operation was necessary from a large sand and gravel aquifer overlain by a thirty foot thick layer of soft organic silt. To build the project without dewatering would have increased its cost by tens of millions of dollars. Dewatering proceeded. In the older section of downtown, where there were frame buildings on shallow footings, significant consolidation occurred but caused little damage. Buildings, pavement and utilities settled uniformly. But in the newly developed sections of the city, where high–rise buildings were founded on caissons through the silts to firm soils, the buildings themselves were undisturbed, but the adjoining utilities, grade slabs and terraces on unsupported ground suffered damage. The state of California accepted responsibility, and awarded contracts to repair the damage. The total cost of repair was less than a million dollars.

Harming the flora

When the water table is lowered for an extended period, (many months or several years), there is concern for the trees and other vegetation in urban parks that make those retreats so delightful. Often concerns are unwarranted; the trees have survived drought in the past. But adequate protection should be afforded, and can be at a minor cost. Two episodes from my experience demonstrate what should and should not be done. During dewatering for the construction of a depressed roadway near the Capitol in Washington D.C. some years ago, concern was expressed for the fine old trees on the mall. Someone in authority ordered the contractor to run a garden hose from his discharge to each nearby tree, and pour water at its base. Fortunately, a botanist happened by and was horrified at what he saw going on. Trees can survive drought he said, but not inundation. The garden hoses were removed. Some years later the "T," Boston's Metro system was building a subway station just a few yards from Harvard Yard, threatening the tall trees shading the statue of the founder of our first university. The engineer in charge of the project wisely commissioned a qualified botanist to monitor the condition of the trees and shrubbery during the two year project. He made inspections and tests monthly; when irrigation was indicated, he provided for it; when there was a need for plant food, he supplied it. It was reported, the foliage was more luxuriant at the end of the project than at its beginning (See Chapter 13 Trees in Urban Construction).

Wetlands

People knowledgeable about such things tell us wetlands form an important part of our ecosystems, and indiscriminate filling of them for development can cause harm. But we are also told that wetlands by their nature undergo periodic changes. They are partly or wholly inundated during rainstorms and they become parched during drought. Along coastal areas a tidal marsh such as we are about to discuss can have its water increase in salinity when high tides and offshore winds bring in sea water. When runoff from the uplands floods the marsh, salinity drops. The flora and fauna that have evolved in such wetlands can survive such changes, or recover after them.

Recently in one of our large coastal cities, dewatering was necessary for a deep sewer tunnel. The alignment was parallel to and about a thousand yards from the beach. It crossed a perennial stream of moderate size, that presented an ideal point for dewatering discharge. However, the specifications forbade it. The author of this requirement apparently had more political influence than understanding of what he was about. It came out in a value engineering meeting that the concern was that the dewatering discharge might upset the salinity of the marsh. The decision had been made without any tests of the salinity of the marsh water, or the groundwater that would be pumped. The contractor's proposal to return many tens of thousands of dollars to the city was rejected, and the unnecessary pipeline to the shore was built.

I once toured parts of the tidal marsh on foot. It sheltered along with its flora and fauna, such things as rusting refrigerators and kitchen stoves, junk automobiles, rotting sofas, worn out tires, and miscellaneous rubbish. It seemed to me that the tens of thousands would have been better spent on cleaning up the trash, and providing footpaths and footbridges so people could enjoy the wetlands resource, while preserving it. But nowadays when logic confronts uninformed ecology, the logic rarely prevails.

Summary

Construction below the water table, particularly in an urban setting, encounters problems. If we investigate carefully the conditions at our worksite, if we familiarize ourselves with the many techniques that might be suitable for solving the problems, we can get the job done at lower cost, with less bother to our neighbors, and with minimum disturbance to the environment. In each generation of engineers and builders, there are a few who want to charge ahead without learning from the mistakes of their predecessors. This is unfortunate. Those who study from past mistakes and then charge, are more likely to be the innovators, to advance technology to ever new heights. To quote Santayana, those who cannot remember the past are condemned to repeat it. In groundwater control, that is costly.

chapter five

Mobilization for a tunnel project in an urban environment

Edward S. Plotkin

Contents

Project mobilization 59
Construction site facilities 60
Project staffing 62
Tunneling in cities 64
Conclusions 67

There are an increasing number of issues that should be considered by owners, designers, and contractors when proposing construction of a tunnel in an urban neighborhood. Ignoring or neglecting to be sensitive to certain critical items has led to negative political reaction, construction delays, budget overruns, and often legal actions.

Project mobilization

Early activities at the commencement of a project are identified as Mobilization. Mobilization by a contractor for the construction of a major project in an urban location should consider four basic objectives: efficient work performance, owner satisfaction, site safety, and minimal community impact. Each of the four objectives can be further reduced to specific activities.

A prudent contractor must be prepared to respond to a multiplicity of new and often unprecedented provisos, usually discovered after the proposal stage, in order to efficiently proceed with large projects in heavily populated

0-8493-7486-3/01/$0.00+$.50

areas. The "old" days of the field office with a superintendent, project engineer and paymaster, are gone. Today, the contractor must prepare his site facility and staff the project in anticipation of the concurrent demands of the neighboring community, various government agencies, labor unions, the project's owner, and the contractor's legal, financial, and insurance interests.

Construction site facilities

The site facility should provide space for contractor's management, administrative, engineering, field labor supervision, and safety personnel. Offices for major subcontractors are often assets since they allow for easy oversight of interrelated activities. It is customary that separate facilities are established for the owner's resident staff, their consultants, and their quality control personnel. This separation of offices strengthens the perception of a formal and professional relationship.

The necessity for field shops for the various labor trades varies, dependent upon the scope and complexity of the project. Mechanic's shops for repair and refurbishing of special equipment, or for the preparation of certain materials will reduce the risk of late deliveries and thereby alleviate potential delays. For example, carpenter shops with lay-down areas for prefabrication of critical items may be economically advantageous. Union requirements may establish needs for particular shops such as electrical, mechanical, and plumbing. Each will incur additional costs for power, ventilation, and telephone.

Warehousing space must be provided for storage of special tools, equipment and parts, electrical and mechanical supplies, and safety related materials. Security measures to avoid pilferage and control for material distribution during the work require a secure location with efficient on-site access. In addition, materials to be delivered and incorporated in the project must be stored in an approved environment. Subcontractors are often paid for materials after they are delivered to the site, and the responsibility for damage incurred subsequently is sometimes difficult to determine.

Most major projects in congested urban locations are restricted in their available storage space. Deliveries of major items should be scheduled for "just-in-time" processing. Utilization of an off-site yard within a reasonable distance provides short-term storage for critical items to reduce the risk of late delivery from non-local suppliers.

A carefully controlled inventory procedure is necessary for the warehousing and storage facilities. Urban sites are attractive to vandals, thieves, and inquisitive children. Site security should be a design factor for the construction facility, including provisions for fire protection and avoidance of unauthorized entry. The contractor is ultimately responsible for injuries to trespassing children. Consideration of the neighboring community and their special concerns is necessary to prevent accidents. Where the local citizens consider English as their second language, the warning signage should be multi-lingual. The contractor must be prepared to explain the basis of his security safety program, particularly during the legal procedures and publicity following an injury claim.

An example of security measures planning involves the stealing of a copper conduit from a NYC Water Tunnel construction materials storage site in a park abutting an economically depressed neighborhood. Since the site was isolated, a solid ten foot high fence surrounding the site was installed with a climbing barrier placed on top. The electric substation for project power contained three independent feeds including transformers, circuit breakers, and network protectors. Part was owned and maintained by the utility company, Con Edison, and part was the responsibility of the contractor. Despite the carefully planned precautions, a total circuit of one of the three feeders was stolen from the storage facility.

It appears that a small (based upon the size of the hole) person burrowed under the fence and unbolted the copper bus bar circuit, including the high voltage protection, while it was live with 13,200 volts. The copper, valued as scrap at $800, was removed through two cut chain link enclosures and a small vent hole in the concrete block of the nearby substation control room. The tools left behind were insulated with rubber tape, not considered sufficient to satisfy a 13.2KV condition. Feeder replacement by the utility company required a team of highly qualified and highly paid technicians to work on the system, since it had to remain "live." The lesson learned is that nothing is impossible when the attraction of perceived value is apparent in an economically depressed location.

Tunnel construction projects require that special amenities be provided. The site facility layout may include a workers' change house and a specific eating area, and provisions for mobile food service. Tunnels usually work on multiple shifts and have on-site eating facilities to avoid the movement of workers from the project site during night shifts. This "convenience" can be a benefit when noise caused by workers who leave to eat in nearby diners in a sleeping neighborhood would have an adverse community response.

Tunnel workers' labor union agreements in several cities stipulate particular conditions for providing services to the miners. In New York City (NYC) for example, a "Hog House" person is required in addition to the tunneling crews, to clean the change house and make the coffee for miners on the job for work breaks and lunch. An eating area in the change house and at work areas in the tunnel must be provided with seating, table, water, and toilet facilities. Of course, other trades that do not have "coffee" provisions in their agreement join the miners. Experience on the NYC Water Tunnel projects provided coffee for 500 workers who averaged four cups during their shift, along with sugar, milk, and cups.

Other examples of special facilities on tunnel project sites are:

- concrete plants
- shotcrete production units
- fabricated steel assembly

- excavated muck storage and handling facilities, and subsequent disposal to truck, train or barge, and material receiving
- water treatment and discharge
- ventilation with appurtenances to control dust and noise
- dewatering plant
- security
- temporary and permanent electric power transformers and substation
- air compressors
- emergency medical station

Each requires planning for their accessibility, power, communications, job needs, staffing, and within an urban setting, their integration into the total program to minimize community impact.

Project staffing

The contractor's staff always includes those personnel necessary for project management, engineering, payroll, cost accounting, scheduling, field supervision, and safety functions. Smaller projects permit individuals to serve with multiple responsibilities. Increased sensitivity to the concerns of the neighboring communities and the growth of governmental agencies involved in permitting and inspections, have increased the need for additional field office staff. The following are examples of added personnel and their duties that also create additional overhead costs and site facility needs.

Community Liaison Officer – to inform those individuals and organizations who are or can be perceived to be affected by the construction activities during the project and after its completion. Local newspapers, bulletins, community cable TV, internet web page, public display of progress charts, meetings with community boards, and other public contact methods should be employed to anticipate or assuage local opposition to the project.

This officer will also record and expedite processing of damage complaints from all outside parties. Heavy equipment operations and blasting activities cause vibrations in nearby structures, often resulting in real and alleged property damage. Proper documentation and routine contact with insurance investigators and government agencies permits timely response and appears to result in reduced or retracted claims. A proactive approach to provide information in anticipation of community actions usually diminishes negative reactions.

In addition to vibration, noise and dust are inherent in construction. These issues must be carefully considered to be minimized to avoid creating a nuisance for abutting properties. There will be complaints! They should be addressed expeditiously and with a level of credibility to maintain a reasonable relationship with the surrounding community.

Labor Specialist – to ascertain and update the status of the various labor agreements with the project workers' trade unions. Duties include the confirmation of the required payments of benefits, annuities, and various stamp funds. Non-payment or false certification of payment by subcontractors can involve the contractor and even the owner in negative proceeding if these things are not verified. The labor unions are required to notify this office regarding member complaints about work rules and conditions. On-site union shop stewards and visiting delegates have a single job contact for labor issues which provides a rapid response to labor problems. On projects without a labor specialist, incidences of employee discontent not receiving a quick resolution or explanation have prompted spiteful phone calls to the US Occupational Safety and Health Administration (OSHA) with punishment, in the form of citations, anticipated for the contractor.

The labor specialist also has the important task of interfacing with minorities, women, and local groups seeking employment. This may not be a contractual requirement, but it is strongly recommended for urban projects. A program to handle minority associations' demands should be planned with input from the owner and local government agencies. The labor specialist will publicize the project employment procedure and record all contacts with individuals and organizations, as well as request labor unions' aid to employ qualified minorities, women, and local personnel. The specialist will also require participation in authorized apprentice and training programs, and maintain records of program attendance, worker attitudes, and work quality, and ask for assistance from the owner, government and police, when groups attempt intimidation tactics. The adverse effect on job cost and the impact on schedules for the project caused by disruption of community groups can be serious and must be avoided.

The aforementioned water tunnel project required negotiation with no fewer than thirteen organizations alleging representation of local minority interests. Among these were such groups as: Black & Latin Economic Survival Society, Harlem Fight Back, Black & Puerto Rican Coalition of Construction Workers, Power at Last, and the Afro-American Coalition of Construction Workers of City. An ability to negotiate rotationally with frequently irrational and occasionally hostile groups, can be developed by prior study. Good preparation often prevents confrontation.

Safety Supervisor – to oversee site safety in general. This area is becoming increasingly important due to rising insurance rates. This office will maintain records of current OSHA regulations and will control the hazardous materials in use at the work site. The safety supervisor will be the contact for the U.S. Labor Department Compliance Officer (OSHA inspector) when the site is visited for random or complaint inspections. The inspector should always be accompanied during the visit. A careful record of any information collected by the OSHA inspector during the visit should be kept for future use in defense of unqualified citations. The safety supervisor should duplicate any photographs taken by the inspector, list the personnel interviewed and, if possible, record any alleged infractions that appear to have been noted. Documentation such as air monitoring is best done before any problem arises.

Electric power transformers must have their PCB content identified, toxic and hazardous ingredients of construction materials should be described in the material data specification sheets (MSDS) and made available to the work force. Particular concern would be lead residue from paint removal activities, noxious fumes created by combinations of construction chemicals, in tunnel work, and methane accumulations at insufficiently ventilated locations. Complaints regarding accidental spills of oils, grouts, and other perceived contaminants, should be processed promptly to avoid litigation or escalation of the incident.

Government Activities Specialist – obtains and maintains current licenses for the project vehicles, trucks, and equipment. Certification of equipment inspections for cranes, hoists, and tanks must be valid to avoid work stoppages when inspected by government agents. This office should also maintain the recent record of licenses for welders, blasters, operators, burners, and any other required government-licensed trades on the project. The license validation schedule does vary by individual and should be kept up-to-date for each trade including each subcontractor's employees.

Tunneling in cities

The construction of tunnels in cities creates major impacts to the neighborhoods along their immediate route. With the growth of population and the resulting need for expanded transportation facilities, new water tunnels, storm water storage and sanitary sewer interceptors, the construction of tunnels has increased dramatically in major cities around the world. The advance of tunneling technology in the past 20 years and the improvements in rock cutter metallurgy, have resulted in the Tunnel Boring Machine (TBM) as the preferred tunneling method. The TBM is utilized wherever the geology and alignment length prove economically advantageous or on projects where the contract documents specify their use for environmental and/or political reasons.

Some tunnels are constructed by cut and cover means. Simply described, a trench is excavated and the tunnel structure is constructed, the structure is buried and the surface restored. In a city environment, the community, traffic, utilities and the quality of life is disturbed. Subterranean tunneling is preferred, with minimal ground openings, if feasible.

The environmental impacts created by TBM mining operations are slight, except for the vibrations caused by the cutters chipping the rock at the tunnel face as the TBM is pushed against the rock, but there are the other community impacts associated with construction. Certain of these cannot be avoided since they are caused by the other surface activities such as rock spoil (muck) disposal, material delivery, site personnel movement, ventilation air currents, etc. There is also a need for drilling and blasting for access shafts and for the approach tunnel in which the TBM will be assembled and launched prior to the beginning of operating the TBM. Drilling and blasting of rock tunnels has a history of several hundred years. About four thousand years ago, tunneling began by heating the rock face with fire and then cracking the rock with cold water. We have advanced. Our present underground infrastructure exists from the past efforts of tunnel workers, miners, and sandhogs working in the hazardous environment, employing explosives to produce the required structures. The new technology of TBM is the alternative construction method and is preferred whenever possible.

Drilling and blasting is economical if the tunnel drive is too short to make the investment in a TBM worthwhile. As noted earlier, a length of tunnel must be developed by blasting for assembly of the TBM. There is no advance of the tunnel during the assembly time for the TBM. The size or shape of the tunnel may also preclude the opportunity to use an existing machine. There are numerous situations in which a contractor has decided to use an existing TBM, which creates a tunnel larger than that required by the contract documents. The result is increased muck disposal and extra concrete necessary to fill the over-excavated cavity. The increased cost must be weighed against the estimated savings associated with the use of a new or re-conditioned TBM.

Scheduling is another important factor when determining the economies of each tunneling method. TBM's usually have a long lead time for manufacturing and delivery. Orders for equipment follow after the award of the contract. The delivery date is a critical point on the construction schedule. Equipment for drilling (jumbos) and blasting activities are more readily available. A jumbo is simply a platform supporting the drills and their backup facilities. On large tunnel or cavern projects, special drill equipment and their support systems may consume some time to design and fabricate, but generally most contractors have some drill machines available which can be quickly adapted. In addition, most drill manufacturers can locate rental or recently used jumbos that can be modified to meet the new project needs.

The possibility of equipment failure is an issue that continually effects the economics of tunneling. The TBM is the critical factor for progress; a major breakdown on a TBM can stop the tunnel advance for an extended period. Tunneling by drill and blast with a jumbo for production of the blast holes avoids the dependence on one piece of equipment. The jumbo usually has several drills which are redundant for the face hole pattern. Since the failure of one drill could slow the hole production operation, the drills are positioned to duplicate coverage of the tunnel face. The drilling task can then continue while a shutdown drill is repaired or replaced.

Another factor which can influence the use of drill and blast is the anticipated ground conditions and their expected influence on tunneling activities. A TBM is designed to operate efficiently within a range of rock hardness. The efficiency of the rock cutters is dependent upon the in situ geology. Where ground conditions vary from very soft to very hard and abrasive, special cutters may be employed, however the progress will be impacted because of this changing environment. Should the mined rock be fragmented allowing loose blocks to dislodge, the machine can be jammed from forward movement. The TBM can be fitted for special rock bolting and roof supporting tasks, but this work also impedes the tunneling progress. Historically, drilling and blasting have progressed through blocky and varied ground conditions by providing the roof support as the tunnel face advances. Progress in difficult ground is reduced but not stopped.

At locations where the tunnel alignment passes above the top of the rock surface and continues into soil or soft ground, only special TBM's are able to operate. The interface of rock and soft ground is often a source of ground water, an issue to be considered in the TBM design.

The cross-section of the tunnel is another consideration in the decision of the best construction equipment. TBM bores are usually circular. This is suitable for water and sewer use. A horseshoe arch shape is more adaptable for railroad use, particularly at passenger stations. The Washington, D.C. subway utilized TBM boring of the running lines, followed by drilling and blasting mining of the station caverns. The bottom section of the bored circle of the running subway lines was concreted to create the flat slab needed for the railroad track. The TBM option for mining many miles of tunnel proved economically and politically correct since blasting vibrations and other environmentally sensitive issues were minimized.

As stated above, the decision to choose between TBM and blast for tunneling is based upon many factors. The engineer may prepare contract documents specifying the TBM method to be responsible to community and environmental considerations. Even if the construction cost is higher than the cost for the blasting method, the contractor must still comply and submit the bid in compliance with the contract requirements. If the choice of a tunneling method is the contractor's, TBM is usually the method of choice when the ground conditions are stable, the tunnel cross-section is uniform, and the tunnel length is long. Choice of the drill and blast method often results when there are shorter lengths, varying tunnel sections, and differing rock qualities.

The public should be aware that even requiring a TBM to mine the tunnel will not preclude the use of explosives on the project. The construction access shaft to the tunnel level activities may require blasting for example. In addition, there will be drill and blast mining of the beginning section of the tunnel to create an area for the assembly of the TBM, often several hundred feet long. The neighborhood surrounding the start of the TBM tunnel construction will be affected similarly to a tunnel totally constructed by drill and blast procedures, and should be so informed.

Conclusions

Construction is a dirty business. When the public hears that the project is a tunnel, they expect to see, hear, and feel nothing. They don't anticipate trucks loaded with muck or deliveries of equipment and tunnel components traveling through their neighborhood, nor the surface facilities inherent with heavy construction. The responsibility for the success of a tunnel project is shared by the owner, engineer, and contractor only with the cooperation from the community, politicians, and the press.

The parties involved must be cognizant of all the issues and must respect the concerns of all impacted groups. Engineers should identify the environmental issues early in the design process and explain to citizens the feasible mitigation measures anticipated. The urbanization of America requires more use of underground space to allow orderly development of cities. The expanded team of owner, engineer, contractor, and the public must cooperate to permit unimpeded progress of the construction of our underground infrastructure.

chapter six

Siting the North River wastewater treatment plant

Nicholas S. Ilijic

Contents

Introduction 69
Historical perspective...70
Progress 71
Conclusions...74
Reference 74

Introduction

New York City, like many other older cities, is continuously faced with the need to operate and maintain existing mature capital facilities. In addition and as a result of the emphasis correctly placed on improving our water environment, new wastewater treatment facilities were also constructed. The following is a discussion of how a state-of-the-art, partially enclosed, fully covered wastewater treatment facility was sited in a congested, socio-economically underprivileged, urban setting. There may very well have been several theoretically valid scenarios, which could have produced a successful conclusion but theory often falls short of success when faced with both the realities and the perceptions of big city life.

Anyone starting an investigation of the feasibility of siting a major facility within a large metropolitan area must at the outset realize that they have taken the first step on the proverbial journey of a thousand miles. The actual time frame will always exceed the anticipated or preliminary schedule by a significant multiplier. This realization is not intended to discourage but rather to prepare the facility's champion to better appreciate the magnitude

0-8493-7486-3/01/$0.00+$.50

of this undertaking and to prepare the team members so that they do not become discouraged with the glacial pace of moving the project forward.

Although this next statement could be considered by some as a lengthy discussion of the obvious, it nevertheless needs to be made. It is always the right thing to do if the proposed project is truly necessary, legally valid, technically and environmentally sound and is sufficiently evaluated. It is also helpful to the success of siting a facility if the consequences of not proceeding with the project far outweigh the negative short-term impacts and substantial costs of proceeding. Once these parameters have been satisfied in the minds of the decision-makers, it is extremely important to identify the champion who will work tirelessly to assure the projects success.

Large complicated projects most often require a multi-disciplined team of professionals who concern themselves with each aspect of the project as it moves forward, but the ultimate success of the project invariably depends on the leadership qualities of its champion. The project's champion does not necessarily have to be the most senior executive or highest ranking official within an organization. However, it does require a person of sufficient rank within the organization to direct and focus the appropriate personnel and expertise that may be required during any given time frame. Above all, champions must generate confidence, be tireless in their efforts, and innovative in both their planning and reactions to an often-changing array of rules, regulations, and oversight personnel. It also helps if the project's champion is viewed by the team as one who will get past the daily emergencies, convince the faint of heart to act and possess that indefinable characteristic of luck.

In order to better understand the various situations that were encountered during the siting of this billion dollar project, it would be beneficial to step back to a point in time when a series of plans were being contemplated to collect and treat raw sewage throughout the city.

Historical perspective

Shortly after the turn of the century, it became evident to the city's public health officials that a master plan should be prepared for collecting, treating, and disposing of the generated wastewater that was being discharged, virtually without any form of treatment, into the surrounding waters. The waters were being polluted badly, and it was apparent that the city was growing at a rapid pace both in Manhattan and in its outer boroughs. This sustained growth would greatly increase pollution particularly at the city's beaches, the upper East River and Lower Bay, as well as the many canals, inlets, and coves that were attracting industry or private housing developments.

To deal with the prospect of a growing population scattered throughout the city, an early plan was presented by the Metropolitan Sewerage Commission (1904-1917), that included 39 separate wastewater treatment facilities, with seven being proposed for the West Side of Manhattan north of Canal Street. One of these seven was located near, what later turned out to be the actual site of the North River Plant. This number was subsequently reduced as it became evident that improving treatment processes and combining facilities to serve geographically larger drainage areas, was a practical and a more cost-effective solution.

In May of 1938, at a significant juncture in the decision-making process, a facility to treat all of the sewage from the west side of Manhattan, from mid-town to Dyckman Street, at a single facility (to be located north of 125th Street and southeast of today's plant site), was advanced by the director of engineering for sewage disposal within the Department of Plant and Structures. In time, this was followed by a review of other appropriate locations along the west side to site a single treatment plant that would provide a preliminary or as later amended a "short period aeration" process. (The city, at that time and well into the 1960's was designing, building, and operating several types of mechanical and biological treatment facilities, e.g., modified and activated aeration, that would not meet the later, 1972-U.S. Environmental Protection Agency (EPA) definition of Secondary Treatment.)

Among other locations, a primary treatment plant was proposed to be located between W. 70th and 72nd Streets. This was a very limited site that required "double-decking" the facility as well as crossing over Riverside Drive and the railroad below. As water pollution treatment requirements changed and both the process times along with the population to be served increased, it became apparent that the W. 70th to 72nd Street site would be inadequate for a plant providing modified treatment.

Progress

Subsequent approval of the current North River site was given in May 1962 by the City Planning Commission, after a public hearing was held on March 28, 1962. This approval was immediately contested and strong community opposition was raised. Approvals by the New York State Department of Health of the preliminary plans followed in September 1962 and November 1963. Art Commission approval was received on December 18, 1963. The U.S. Army Corps of Engineers approved the outfall for the plant and issued a permit for construction in December 1963. The New York City Site Selection Board approved the site on February 17, 1964 and, on May 20, 1964, the Mayor signed the certificate authorizing the Corporation Counsel to acquire property from New York Central Railroad to accommodate the North River Treatment Plant.

By 1967, a defining resolution was promulgated through the Hudson River Compact which determined that secondary treatment, e.g., better than 85% removal of the biochemical oxygen demand (BOD) and suspended solids (SS), must be provided at any treatment facility discharging into the Hudson. This clarifying resolution required an indepth study to be performed in order to determine the suitability of the current location for a much larger facility than was previously envisioned for what was now commonly known as the North River Treatment Plant.

Many elected and appointed officials also voiced objections to the site during this period; included among them was a future mayor of New York City. As each phase of the project moved forward, including the mining of an intercepting sewer well below the surface of Riverside Park, it was met by efforts to halt or delay the start of construction. The greater focus of concern for the residents of the upper West Side Harlem community was the perceived adverse health effects, plant odors, and the magnitude of the plant and its aesthetic appearance. To effectively deal with the magnitude and aesthetics of the proposed facility, a number of well-known architectural firms were hired in turn, to develop a plan to mitigate these impacts as well as enhance its utilization for the community. The community also took strong exception to the city's intended use of ozone to treat the air before discharging it to the atmosphere. This process was changed to one utilizing carbon filters and liquid scrubbers, as a result of the public's concern.[1]

The city commissioned an additional study in 1968 which concluded that a secondary treatment plant capable of processing the estimated flow could be located on city owned or acquired property between W. 137th and 143rd Streets along the Hudson River. This property was zoned for manufacturing and could, after receiving a special permit by the City Planning Commission, be used to site a wastewater treatment facility. Most of the property was under water and required that the 28-acre deck to be built over the river and supported by approximately 2305 concrete-filled steel caissons. The majority of which were 42 inches in diameter, 200 feet in length, filled with concrete and socketed into rock.

In order to provide the required treatment, the study's conclusion recommended a "double decked" facility with five 30 foot deep aeration tanks (normally15 feet); in addition, the chlorine contact tanks would be located; below the final settling tanks. (The EPA, after reviewing flow projection figures determined that the plant should have a permitted dry weather capacity of only 170 mgd and not the 220 mgd or as a compromise, 190 mgd that the city suggested). As a result, the final design was further modified to accommodate this change by reducing the pumping capacity and removing one of the 30 foot deep aeration tanks. The cleared space was utilized for the construction of an amphitheater, as part of the proposed Riverbank State Park.

It was also at a critical point in time and in order to ensure the success of the project that the then Governor of New York State, Nelson Rockefeller, promised the West Harlem Community that a "meaningful state park" would be constructed on the roof of the North River Treatment Plant. Subsequently, the plant was designed and redesigned several times. It changed from a facility that resembled an industrial grouping of several buildings, to a completely enclosed windowless building. One plan called for the entire facility to be covered by what some community residents described as a "moonscape" with reflecting pools and fountains of various sizes, some capable of sending jets of water a hundred feet or more into the air. One design included a magnificent central bridge-plaza crossing from Riverside

Drive over the existing Penn-Central Railroad Yards and the Henry Hudson Parkway. It became evident that an in-depth study of the community's needs for recreational and cultural facilities and appropriate access must be coupled with the state's ability to fund a "meaningful park."

While the plant was in final design and the 28 acre foundation already under construction, an oil crisis forced the cost of energy to reach an all-time high causing the elimination of the totally enclosed, window-less exterior in preference to a partially opened one. The roof, which was capable of supporting a combined dead and live load of over 700 pounds per square foot, would remain, but approximately 60% of the walls would be opened. This modification allows for natural ventilation and significant savings in both the capital and energy costs of moving and controlling the temperature of millions of cubic feet of air.

As the project progressed, there was a continuous stream of issues requiring quick and often very innovative actions before the first caisson for the deck foundation was driven:

- City Planning Commission – Special Permit
- Con Edison Power Transformer Station (accessible from a mapped street)
- Fire Department Access (requiring a legal street of sufficient width in order to accommodate fire apparatus)
- Creation of New Street (mapped from 135th Street to the plant's entrance)
- Purchase of land for the multiple Wick's Law contractors' use, along with an access road to the site, from the bankrupt New York Central Railroad
- Board of Estimate Approval to award the $228 Mil Foundation (Deck) Contract (at the time it was the largest single, non-defense contract awarded in the Western Hemisphere)
- Significant minority business involvement and employment of local construction personnel
- A complete reassessment of how to prepare future wastewater plant construction cost estimates
- A series of court mandated milestone dates for each phase of the designed construction

Each of these issues in turn contributed to the almost daily emergencies being dealt with by the champion and the team members, in order to continue progressing the project to its completion. The foundation (deck) alone required a team of five major construction contractors. They in turn required 39 insurance companies to fully bond the project.

A better understanding and appreciation of the actual construction of this monumental treatment facility will require a more comprehensive dissertation along with a site visit.

Conclusions

None of the 14 modern wastewater treatment plants operated by the city are ever really completed. Their combined ability to successfully treat more than 1.5 billion gallons of dry weather flow per day requires continuous, diligent maintenance and periodic reconstruction or upgrading of specific structural and process features. North River is no exception and after a rocky start and a significant amount of effort on the part of dedicated engineers, administrators, and the community, additional millions more in capital dollars have been spent in striving towards becoming a more neighborhood, accepted facility.

The local community, along with the greater regional community, has benefited greatly from the significant mix of activities provided by Riverbank State Park, which occupies the entire roof of the North River Treatment Plant. One could view the treatment plant below as a functional pedestal supporting this meaningful park.

Major capital projects will continue to be sited in urban areas. Their success will depend as much on each of the many individuals that represent the owner, engineer, constructor, and community quickly understanding each other's goals for the actual design or construction methods proposed.

Large projects sited adjacent to established communities require innovative concepts and substantial funding in order to provide the affected community with a positive betterment. That betterment is best determined by the needs and wishes of the local residents. A continued exchange of information among the owners, constructors, and operators of the facility to better understand the concerns of the local community is always necessary to insure that the facility will be operated well, adequately maintained, and will continue to strive to be a good neighbor.

Reference

For the most part, the city depends on a collection system that combines raw sewage with storm water and discharges the mixture through hundreds of combined sewer outfalls (approximately 450). Secondary treatment plants employing the step-aeration/activated sludge process were to be hydraulically designed to treat up to twice the mean dry weather flow (2 x MDWF) through the primary settling tanks and up to one-and-one-half times the mean dry weather flow (1-1/2 x MDWF) through the aeration and secondary settling tanks (final clarifiers). This prudent capacity requirement along with the quantified definitions (NYSDEC permit requirements) for Secondary Treatment effectively dictated the physical size of the future EPA "permitted" wastewater treatment facilities throughout the city. The other very real constraint impacting the siting of new facilities or the expansion of existing facilities was the limited acreage available within the city and the city's rigid zoning requirements. The North River Plant is located in a M1 Zone.

The North River Drainage Area is bounded on the south by Banks Street in Greenwich Village and extends along the entire West Side to the northern tip of Manhattan, including a stretch along the Harlem River to just south of W. 201st Street. Its eastern boundary approximates the north-south centerline of Manhattan Island. This area is home to over half a million residents and a sizeable number of daily commuters. The estimated hydraulic capacity for the facility (Design Year 2005) was determined by USEPA to be 170 million gallons a day of dry weather flow and twice that amount as wet weather flow, to be treated prior to discharge. This 28 acre facility did not displace a single person or business.

Note: For a period of time during the exploration of North America, the Hudson River was known as the North River, while the Delaware River was known as the South River. As a result, our Hudson River piers are called North River piers leading to the historically appropriate naming for this wastewater treatment facility.

chapter seven

Getting along with the existing infrastructure

Reuben Samuels

Contents

Introduction 77
At grade roadways 78
Construction over city streets 78
Construction under city streets 79
Existing utilities 79
Building foundations 80
Open cut subway construction 81
Elevated structures or tunnels – is there a choice? 82

Introduction

Getting along with the existing environment is a particularly complex problem in urban construction. The range of "getting along" varies from a simple barricaded street opening for utility work to an open cut subway project constructed under decking carrying both pedestrian and vehicular traffic as well as work inside operating rail terminals without disruption of train operations. The open cut subway projects in Los Angeles, San Francisco, Atlanta, WMATA, New York City, and Boston have left many years of bruised neighborhoods and lawsuits. In the historical perspective, most completed subway construction has enhanced neighborhoods and raised property values. The anguish of the abutters during construction is sometimes exacerbated by news media and/or politicians.

0-8493-7486-3/01/$0.00+$.50

For this chapter, the topic has been subdivided into seven categories ranging from at grade roadways to complex open cut subways and concluding with a discussion, from an impact on neighborhood point of view, of elevated structures versus underground structures.

At grade roadways

For purposes of discussion, this category is the replacement/reconfiguration of multi-lane highways while carrying the existing traffic volume. Clearly, a full or partial detour could be a solution, perhaps having a faster construction schedule — but the essence of a detour is not a trade off — rather two neighborhoods are adversely affected. The current trend seems to be a combination of:

- Restriping for narrower traffic lanes.
- Creating additional lanes from restriping and impinging on medians and shoulders.
- Limiting work areas to $^1/_4$, $^1/_3$, or $^1/_2$ the number of lanes with night work a contract obligation.

In some major cities and many smaller geographic entities, light rail, with a minimum of depressed or elevated sections (modern adoption of "trolley" lines) has proven to be the most economical solution for an effective mass transit system. As a matter of course, this construction along existing streets will be the cause of disruption, but an order of magnitude less than the open cut subway construction.

Construction over city streets

Typically, in an urban setting, both vehicular (e.g., Gowanus Expressway in Brooklyn, New York, and Central Artery in Boston) and light rail exist on elevated structures. Some early twentieth century elevated light rail structures were utilized for portions of new elevated vehicular structures. The typical structure of painted steel and concrete pavement on roads required a high degree of maintenance which was not always part of an ongoing program. As a result of a decaying infrastructure in general, much of this type of structure needs a full replacement and or very specific enhancement to reflect additional traffic and new general safety criteria. There are many cases of structures being erected over urban expressways, especially along riverfrontage. Examples of this can be seen at the East River in Manhattan where many hospitals fronting on the East River Drive and now, by dint of supporting piers on the land side and water side of the drive, multi-story hospital buildings are now in place which are built up from steel trusses which span the highly trafficked East River Drive.

Construction under city streets

In the late nineteenth century, the elevated railroad structures began to be replaced by tunnels under the street right of way. The surging ridership in New York City provided incentive to construct from the original 20 mile length in 1904, to hundreds of miles in the present day, both in open cut/cut and cover as well as bored tunnel construction. In addition, vehicular tunnels began to be built from underpasses, through many block long tunnels to a complete underground ring road as in Singapore. The necessary entrance and exit ramps required in an urban setting require extensive and expensive supplemental structures to both underground and elevated vehicular structures. Other types of tunnels under city streets are for water or sewage hundreds of feet below the street surface, e.g., 24 foot diameter water tunnels in New York City and 30+ foot diameter drainage and reservoir tunnels in Chicago. In New York City, there is a 2 block long tunnel to transfer mail between the main post office and an annex.

Existing utilities

An urban street can have a veritable maze of utilities serving the immediate neighborhoods as well as parts of a larger grid system that may have dimensions expressed in miles. A typical avenue in New York City (100 foot building line to building line) might have:

- Electric power, both primary and secondary voltage on both sides of the avenue.
- Gas mains on both sides of the avenue.
- Water mains on both sides of the avenue.
- Sewers on both sides of the avenue.
- Storm drains on both sides of the avenue.
- Telephone lines, both copper (being phased out) and fiber-optic on both sides of the avenue.
- Steam lines (at 400 psi).
- Telecommunications and data fiber-optic systems.
- Cable TV.
- Deeper collector sewer and major water lines to feed and drain the smaller pipes.
- Services to connect street mains to individual buildings.

This stratum of utility maze can be 5 to 20 feet below the street surface and represent a major contingency (the existence and/or location cannot always be known) in open cut excavation. Open cut excavation operations must consider maintaining utilities in place, or if there is a direct foul between new construction and existing utilities, the utilities must be taken out of service or relocated – temporarily or permanently.

Building foundations

The foundation, extending from street level to the lowest basement level, including foundation elements to rock or other bearing stratum, while a small percentage of the total building cost is a major and critical part of the schedule. Often, to save time, the foundations are started knowing only the footprint, depth and location of column contours. Permits must be obtained for:

- Erecting perimeter fences.
- Pedestrian passageway outside the fence.
- Storage of materials and equipment in the curb lane.
- Maintenance of vehicular traffic, possibly closing additional lanes in off-hours.
- Storage of and use of dynamite.
- Use of water from fire hydrants.
- A sidewalk crossing permit (for entrance and exit from the excavation site).
- Expanded hours of work, if restricted (e.g., no blasting after 7 p.m.).

Detailed surveys must be performed for the site with specific information (plan location and elevations) provided for adjoining structures. The location of the actual existing structure (vis-à-vis) the property line — whether the adjoining structure is on the property line, behind the property line or encroaching into the new building construction is extremely important legally and in regard to the title, guarantees/financing arrangement. Borings should be carried to below the excavation subgrade and/or to a depth in soil that is not underlain by soft or weak soils and/or to rock (with 10 feet or core) if rock is below subgrade. Ground water levels should be taken. Sequentially, the following operations would take place:

- Commence support of excavation, typically soldier beams and lagging.
- Expose adjoining building foundations: – if on rock, commence line drilling using a three hole per foot patterns; if adjoining building foundations are not on rock, underpin to top of rock or three feet below new excavation subgrade or new footing bottoms.
- Obtain magazine-blasting permit(s); review methods and procedures.
- Support of excavation at the perimeter (e.g., soldier beam and lagging) for adjoining streets(s) and close by utilities, at adjoining building(s) – stabilize existing building walls – if not on rock, underpin to rock or 3 feet below lowest excavation subgrade, if not underlain by soft or weak soil; where perimeter rock faces are created, the use of rock bolts and shotcrete may be used to stabilize potential slides and concrete can be used to fill cavity/spaces resulting from fallout from the rock face.

- General excavation to the bottom of the proposed foundation; depending on whether rock exists at subgrade, shallow depth below subrade or large depth below subgrade the footings can bear on rock soil or rock or deep foundation to rock [i.e., piers, piles, LBEs (load bearing element-slurry construction].
- Concrete footings and/or pile caps poured as footings, both spread on soil and bearing on rock, as well as deep foundation elements are completed.
- Perimeter and interior change in grade walls poured, so the bathtub (perimeter wells, slab on grade and footings and/or pile caps), is completed and ready for steel or concrete superstructure.
- Restoration of sidewalks, paving, and utilities are performed.

Open cut subway construction

A boring program for soil, rock, and ground water levels should be instituted along the alignment with some borings carried to rock as the depth, below excavation subgrade to top of rock is established; soft or weak soils and existing ground water levels should be identified. Sequence of operations would be:

- Appropriate permits, traffic patterns, and hours of work limitations are established.
- Vehicular and pedestrian traffic are maintained or limited.
- Existing utilities are exposed/explored for maintaining in place, supporting (off decking beams) and/or relocating.
- Support of excavation is installed, typically soldier beams and lagging to rock above subgrade or a minimum of 3 feet below excavation subgrade with soldiers penetrating 10 feet below excavation subgrade.
- Decking is installed (e.g., 36 WF beams on 10 foot centers with precast decking spanning 10 feet supported on beams which bear on individual soldiers or on a cap beam on top of soldiers).
- Dewatering system is installed.
- If hazardous materials exist, suitable removal and disposal is arranged – it is more efficient in time and money if this cost is assumed by the owner (note legal interpretation of "the owner owns the ground").
- Excavation is done with bracing and/or tiebacks installed as the depth of excavation advances.
- Subgrade is prepared for either a soil or rock footing or, if rock or good ground is below subgrade, then foundation elements are installed (piers, piles, LBEs).
- Subgrade is prepared with filter fabric, drain system and granular fill.
- Invert walls and roof of subway structure are installed.
- Backfill is done.
- Restoration of sidewalks, paving, and utilities are performed.

Elevated structures or tunnels – is there a choice?

As we look back and plan ahead, it becomes apparent that two divergent conclusions are upon us. One is that the at grade railroad construction commencing in the nineteenth century served to divide the town and city into "good and bad sides" of the towns and cities. As commuter rail transportation evolved, most large cities progressed through elevated railroad systems in the central business district to rail transportation going underground (usually in the right of way under streets and avenues) in the early twentieth century. The New York City subway system started at the beginning of the twentieth century, followed very quickly by the development of Penn Station with a completely underground tunnel system from New Jersey, through Manhattan, to the borough of Queens and, in the same time period, the Grand Central Terminal was constructed with tunneling (cut and cover) up to 99th Street and Park Avenue. It should be noted the tunnels from Grand Central Terminal were constructed with cut and cover tunneling to 99th and Park Avenue and then became elevated. In hindsight, one can only speculate as to the socioeconomic impact on neighborhoods if the 40 blocks of elevated railroad were in tunnels rather than on elevated structures.

The other conclusion might be that underground space is good, in fact, much better than major highways at grade or elevated structures (many are still in operation). In looking at existing early 1900 commuter rail elevated structures and elevated highway structures (most have been built as a result of the nationwide highway network started in the 1950s), all are operating at multiples of the design capacity (two specific cases are the Gowanus Expressway in Brooklyn, New York and the Central Artery in Boston).

As underground space construction became more cost effective, especially in the "good ground" such as the granites of Norway and Sweden, the use of underground space clearly made available to communities more environmentally better space than above ground space.

The overall concept of "a sustainable environment," especially when applied to the reality of dozens of new 10,000,000 population cities, many in Third World countries, provide a huge collection of engineering/socioeconomic problems but without creating the huge required funding sources.

chapter eight

Exterior wall renovation in urban areas

Gregory T. Waugh

Contents

Six Penn Center, Philadelphia 84
1100 Avenue of the Americas 86
660 Madison Avenue (Barney's) 89

This chapter describes three recladding projects as examples of the challenges of exterior wall refurbishing and replacement in urban areas. The projects are Six Penn Center, (Philadelphia); 1100 Avenue of the Americas (New York City); and 660 Madison Avenue (New York City). Although each project was different from the others, certain elements remained constant. These include bringing each up to today's standards for energy efficiency and temperature control, accessibility to the handicapped, elevator service, air and water infiltration, servicing and maintenance, security and communications, as well as such basics as bringing each up to current code standards and repairing any structural deficiencies.

Rehabilitation may also be chosen over new construction when there is a need to upgrade the building's image and marketability at the same time that existing tenants with long leases (who are unwilling to move out) are accommodated. Sometimes the amount of work on an income-producing property necessary to bring it to a competitive posture does not warrant its complete replacement.

Three renovation projects will be examined to illustrate the kinds of issues and problems which may be encountered in the major overhaul of a large commercial/institutional property in an urban area.

0-8493-7486-3/01/$0.00+$.50

Six Penn Center, Philadelphia

The first example is Six Penn Center in Philadelphia (Figure 8.1). This was the corporate headquarters building for Conrail that was built in 1955 but which has been unoccupied since 1992. It is an 18-story steel framed structure designed by Vincent G. Kling & Associates and was part of a transportation center block, which incorporated a bus terminal and garage. Its outstanding location at Market Street and 17th Street in Center City, next to, and on the same block as The Mellon Bank Building, designed by Kohn Pedersen Fox Associates PC (KPF), made it a prime candidate for development. However, its exterior skin, consisting of a simple limestone with strip, single–glazed windows on all sides, was both dated looking and suffering from a variety of technical problems.

Figure 8.1 Six Penn Center

When the owner, The Rubin Organization/Equitable, first asked the architect, KPF to make recommendations for upgrading this property for a potential tenant, Morgan Lewis Bockius, the focus was on upgrading the existing facade, re-modeling the lobby and office level toilet rooms, replacing the elevator cabs, and making other cosmetic and technical improvements. A partner in the property parkway, was planning an exterior ramp structure on the west façade for the length of the building. The first design was to move the ramp out from the mass of the building to create a reveal expression to emphasize this new element and anchor the building. The reveal space between the ramp structure and building created a transition element and space for the parking exhaust louvers, more parking, and a larger space for the ground floor lobby.

Since the façade was very dated, several studies were conducted to replace the exterior wall in part or in total. After several recladding schemes were priced with a construction manager, it was determined to maintain the limestone which was in good condition, and remove limestone panels between the punched windows to create a continuous ribbon window. The first step in evaluating the wall was to remove a portion of the wall with the window wall consultant, construction manager, engineers, and KPF present. Since the wall appeared very dry despite the single line of sealant protection and minimal steel window frame, it was decided to proceed with the existing ribbon window scheme. After an analysis of the wall back up, and removing the portions of wall between the punched windows, it was determined that a continuous sill bent plate reinforcement was required to resist overturning moment of the wall. Also diagonal steel braces were required at the heads of the window for the same reason. Fortunately at the head of windows, a continuous lintel angle was used instead of single lintels for each window. The fire alarm system was replaced, sprinklers added, and the HVAC and electrical and plumbing systems partially reused. The air ducts at the columns that once served air induction units were enclosed in fire–rated enclosures with fire dampers to serve new VAV boxes in the ceiling.

The parking ramp structure façade was studied in conjunction with a through block plaza connection that opened up to the Mellon Bank plaza. Articulated stainless steel panels were used on the ramp facade and tied into the ribbon window stainless steel frame and column covers. A gutter at the head of the window was designed for controlled water leakage and new thermal insulation with foil facing was used on the interior of the wall. Insulating, energy efficient low 'E' glass was used in conjunction with the taped foil face insulation to create an air and watertight wall for maximum comfort of the occupants, and to relate well to neighboring buildings, ASTM field tests were conducted at various stages to confirm water tightness of the wall.

In summary, an empty, dated and dreary building was brought back to the marketplace and repositioned to again be a valuable corporate address that completes the composition of the entire block by tying materials as stone and stainless steel to the existing Mellon Center which creates its own distinctive presence. Careful analysis by the design/construction team permitted the owner/partnership to make informed decisions on the most cost-effective manner of reclaiming the property and creating a valuable investment.

1100 Avenue of the Americas

The second example is 1100 Avenue of the Americas, located at the corner of 42nd Street. The original structure was actually two distinctly different buildings, one located on top of the other. The first portion was built in 1906 and was a seven-story office block built using cast-iron columns and steel beams supporting cinder-fill floors formed over flat terra cotta arches. The columns' base plates rested on footings made of two stacked layers of steel grillage resting on bedrock. In 1926, an additional 8 stories were added directly on top of the existing structure. This structure comprised steel columns and beams supporting floors of poured gypsum reinforced with wire cables hung in catenaries between the beams. The gypsum was obviously chosen to minimize the weight of the new construction. The facade of the 1906 building was typical of the period with terra-cotta swags, medallions and cornices embellishing every level while the 1926 addition was a much plainer brick box.

Over the course of the years, the character of the surrounding neighborhood had changed dramatically. 42nd Street became synonymous with video shops and cheap entertainment. However, Avenue of the Americas remained a prime location for offices and is home to many Fortune 500 companies, although at the time the renovation was being planned, the best properties were still located several blocks to the north (Figure 8.2).

The owner's goal was to vacate the building of its somewhat difficult tenants and to modernize it in order to attract the sort of companies who would pay top rents. In order to do that, it was necessary to combine a Class A physical plant with a desirable address and then to provide some additional features which would attract tenants to the southern fringe of the choice mid-town area. One advantage was the large 20,000 square foot floor plates which rise undiminished in size from ground up to the top (15th floor) of the building without any set-backs. By renovating the old structure, it was possible to maintain the existing floor dimension (which were actually increased in size by filling in the old light and air shafts). Had the old building been torn down and replaced with a new one, current zoning requirements would have dictated 20 foot deep set-backs on both 42nd Street and the Avenue of the Americas, resulting in smaller floors, less flexible for modern office planning. This was the major impetus for rehabilitation versus new construction. Another factor favoring the rehabilitation (rehab) option was the availability at the time of the Investment Tax Credit, which allowed a substantial subsidy for a project which complied with certain criteria, one

Figure 8.2 1100 Avenue of the Americas — Before.

being that to qualify, a rehab had to maintain at least 75% of its existing exterior envelope or to replace at least that amount of the old envelope with new construction in the same location as the old "curtain" and supported by the same structural framework that supported the old envelope.

Although the building is located well to the east of the less desirable stretches of 42nd Street, the owner decided to relocate the main entrance away from 42nd Street to the side of the building facing Avenue of the Americas. This was done in order to avoid the stigma of a W. 42nd Street address even though this meant a much larger lobby than would have been necessary had the entrance remained on 42nd Street. Having now achieved a desirable address and provided a feature that could give a marketing edge to the property, the next goal was the total modernization of a building whose mechanical and electrical systems, elevators, restrooms, and lobby

Figure 8.3 1100 Avenue of the Americas — after.

were all obsolete. Also, it was felt that the image of the facade would have to be made to look as if the project were completely brand new (Figure 8.3).

In sum, the program required that the existing building be stripped down so that only the existing columns, beams and floor slabs remained. Even the fire stairs were ripped out to allow a complete re-organization of the cores. One of the main challenges of the renovation was to reinforce the existing columns to take the loads of the new slabs in the old light courts on the north side of the building. The structural engineer rejected the notion of bolting reinforcing plates to these cast iron elements because of their brittleness and the associated risk of their cracking if they were drilled into. Welding new members on was also ruled out due to the difficulty of welding steel to cast iron. The solution was to build a new, completely independent column around the old one with no attachments whatsoever joining the two together. These new columns were built up of 4 vertical 8″ × 8″ × 1″ angles forming a square box around the old column. These angles are laced together by a lattice composed of 3″ wide steel diagonals made from flat 1/2″ thick steel bars bolted to the 8″ angles. A completely independent footing for each "cage" column was also necessary. To achieve this, the old footings were exposed by carefully picking away the existing concrete covering the grillage and then sandblasting the grillage to obtain good bonding with the new poured concrete footing which encases the grillage and serves to support the new cage columns.

Another significant structural problem was the loss of a considerable amount of rigidity in the building frame due to the removal of the massive masonry exterior wall. Although the exterior wall was non-load bearing, it did provide much of the lateral resistance to wind loads. Four new steel wind trusses were introduced in the core, each attached to two of the new steel cage columns. Details for creating new moment connections where existing steel beams meet new cage columns were also developed.

Once the new structural work was in place, many aspects of the project became not too different from conventional new construction. Since absolutely nothing of the old building other than the columns and floors was being retained, the installation of the new HVAC (including four-pipe fan-coils at the perimeters), electrical, elevators, etc. was all fairly straightforward. Much of the complexity of the project lay in dealing with the impact of the existing, irregular structural grid on the layouts of the cores and the fenestration. Another area of complication was the attachment of the new glass and metal curtain wall to the old floor slab. Although weight was not a factor in this instance, the thickness of the new wall was far less than that of the masonry wall it was replacing. Therefore, several details for extending the old slabs had to be developed. The interfacing of the two different existing floor types (terra-cotta and gypsum) and the varying dimensions of the existing spandrel beams from the new face of building was challenging for the designers.

660 Madison Avenue (Barney's)

In contrast to the complete removal and replacement of all building elements other than the basic structure, the project at 660 Madison for Barney's New York illustrates a situation in which parts of various systems were worth saving. This introduces a whole different level of complication in that careful analysis and testing is required to determine which components can be re-used and which must be replaced. 660 Madison Avenue is a steel framed 22-story office building put up in 1958 to the design of Emery Roth and Sons (Figure 8.4). Its outstanding location at the junction of the mid-town commercial district and the elegant residential and retail Upper East Side along with excellent management has made this a very successful property. However, its exterior skin, consisting of a simple glass and metal panel curtain wall on East 61st Street and Madison Avenue and a brick and strip window

Figure 8.4 660 Madison — before.

wall on East 60th Street and the lot lines, was both dated looking and suffering from a variety of technical problems. When the architect was first asked by the owner, Metropolitan Life Insurance Co. (Met Life), to make recommendations for upgrading this property, the focus was on replacing or re-covering the existing facade, re-modeling the lobby and office level toilet rooms, replacing the elevator cabs and making other cosmetic and technical improvements. It was felt at that time that the existing HVAC and electric systems were fundamentally sound. The building was also in compliance with all mandated life-safety requirements.

Before studies had proceeded very far, however, an entirely new element was introduced in that Barney's New York, a high-end clothing store for men and women, and Met Life joined forces to form a condominium, with Barney's purchasing the lower nine floors plus the cellar for the purpose of fitting out these premises as their new flagship retail facility. The conversion of seven floors of the building from office use to retail immediately mandated the major upsizing of the building's HVAC plant as well as the provision of three additional truck berths to handle the increase in delivery traffic. While the new program still called for replacement of the existing facades, each owner wanted a distinct identity for his respective portion of the building, expressive of the very different functions located within. This completely changed the program.

In addition to the changes noted above, other modifications to the existing building were incorporated (Figure 8.5). These include the addition of a smoke purge system, the extension of various elevator shafts both up and

Figure 8.5 660 Madison — after.

down, a new computerized control system for the existing elevators, new escalators to connect eight floors of retail, a new fire alarm system, an independent sprinkler system to serve the retail areas, the creation of new building set-backs to help articulate the massing of the retail unit, relocation of various stair shafts, etc. Of course, each change required that the existing structure be checked and reinforced as necessary to accommodate the new loadings and configurations. Despite the extent of the changes, it was found that various elements of the existing building could still be kept in place and re-used.

Cost and feasibility studies were conducted of the exterior wall to determine the most efficient wall system possible. The massing of the building was also studied to improve the 60's wedding cake setbacks and provide a more appealing configuration. An overcladding system was studied which maintained the existing aluminum framing but added new glass and trim. A modified overcladding system was studied maintaining some existing elements, and total replacement of the wall. It was determined that the total replacement was the most efficient system due to the increased ease of adding new panels without the logistical problems of dealing with existing elements that may be out of plumb, etc. The buildings exterior wall was stripped to the structural system and a new exterior wall system consisting of precast concrete panels with wrought iron railing detailing and inset aluminum windows was designed. The panels spanned from column to column as not to place loads on the edges of slabs which would not take the load without significant reinforcing. The panels were tested in a full size mock-up in Florida to check the performance criteria of the window panels.

These two projects illustrate that the renovation of an existing property can often be the most desirable solution to a development opportunity even when very extensive changes are involved. This is true financially, but also in terms of enhancing the urban fabric. Once the decision has been made to renovate, it is of fundamental importance to identify and understand the physical properties and condition of those elements, which are to be retained. Armed with this knowledge, the design team can then proceed to determine how best to integrate the new elements with the old.

chapter nine

Community relations and urban design: the New York Psychiatric Institute case study

Jill N. Lerner

Contents

The institute..94
The neighborhood..95
The dilemma..97
The strategy emerges..98
Plusses and minuses..99
The design role...100
The site and program...100
The presentations...102
A positive response..103
Epilogue..104

In the late 1980s, the New York State Psychiatric Institute (NYSPI), a public institution dedicated to psychiatric research, considered their need to renovate and expand their existing 60-year old building in upper Manhattan. After several years of study, it became apparent that expansion in place was impossible, and that renovation could not be accomplished while keeping the institution up and running. The institute looked for reasonable sites in the neighborhood, considering any feasible method of expansion — renovation, new construction, or even total relocation — without success.

0-8493-7486-3/01/$0.00+$.50

By 1991, the Institute had approvals and funding in hand to begin the design of a new facility on a questionable site nearby. The New York City architectural firm of Ellerbe Becket, Inc. was selected to design the new, state-of-the-art psychiatric research facility, under the leadership of designer Peter Pran and myself. However, it was not clear that one could actually construct a building on the site proposed. Despite its private ownership, the chosen site had been "mapped" by New York City as parkland, a highly prized commodity in this dense, urban environment. It was evident that more than architectural skill would be necessary to get anything built.

Located in the Washington Heights area of Manhattan's upper west side, it was well known at the start that this would be a controversial project. As a positive contribution to the neighborhood, an employer of local residents and a nationally renowned research institute, NYSPI had a lot of supporters. As with many New York City neighborhoods, we all knew we could count on a vocal community debate. Certain factions were sure to oppose the project, and could even stop the project from proceeding at all. Although we embarked upon a carefully planned political process, there was, perhaps, a 50% chance of success as the project began.

As with all complex problems, the background and setting of the situation were critical to understanding the solution and process that could lead to success. Timing, history, leadership, creativity, and strategy all played key roles. The carefully crafted presentation process that emerged built consensus among diverse supporters, and kept objectors to a minimum. Numerous groups that could not support the project agreed to remain silent, rather than fanning the flames in vocal public debate; or worse, in filing lawsuits that would have surely delayed and even defeated the project.

After an 18-month process including 52 community meetings, multiple public hearings, and numerous design changes the project went forward and was completed in 1998. Today, the institute is thriving beyond expectations, in an award-winning building that has improved the Institute, the neighborhood and even the skyline of New York. In the end, a determined combination of a thoughtful strategy, respect for neighborhood input, intelligence, creativity, and positive determination allowed the project to go forward and the Institute to succeed in its mission.

The institute

The New York State Psychiatric Institute was founded in 1896, and is the oldest psychiatric research institute in the country. Many important medical discoveries were made at the institute, and almost 100 years later, it continues as a leader in research grants with impressive accomplishments in psychiatric research and treatment. It is a major force in the field of psychiatric research containing 25% of the fully funded psychiatric research beds in the country, and the largest grant-supported department within the university's medical center.

NYSPI is unique in its public and private affiliations. It is a public institution, a part of the New York State Office of Mental Health (OMH) based in Albany, and one of two major psychiatric research centers in the OMH system. It is also the headquarters for the Department of Psychiatry of Columbia University's Medical School. While the operating budget for the institute is funded by state money, private grants support much of the research activity contained within. The institute's director, Dr. John Oldham, is firmly entrenched in both camps, as Chief Medical Officer for the New York State Office of Mental Health, and as a longstanding faculty member at the medical school.

The institute building had become outdated, from an architectural and engineering perspective (Figure 9.1). NYSPI had outgrown its original building, a main structure constructed in 1929, and a more modern building, the Kolb Annex was built in the 1970s. Additional parts of the institute were housed in rental space. The main building was built for NYSPI when it moved from Ward's Island in the 1920s to join the new Columbia Presbyterian Medical Center complex. By 1991, this confusing building, with its entrance on the 10th Floor contained areas, 60-year old patients labs, and offices. The Kolb Annex, although 12 stories in height, was designed as a true "annex" from a mechanical and systems point of view, dependent on the main building for services, loading dock, etc. Any new replacement building would have to be physically linked to the Kolb Annex, as its umbilical cord and lifeline.

It was clear that for NYSPI to remain a leader in psychiatric research, a new, modern facility to replace the main building was critical. Without it, the survival of the Institute was truly at stake. It was only a matter of time until the building systems would be unrepairable, until dismal and non code-compliant patient areas would be unusable, and, most importantly, until the institute would no longer be able to recruit the leading researchers, or "principal investigators" that kept the NYSPI on the cutting edge. Not only did the institute need a new building, it needed one commensurate with the prominence of the institute itself. In this competitive research environment, with other institutions building their own new facilities, it was either "keep up or close."

The neighborhood

Washington Heights in upper Manhattan is a dense area, primarily a residential neighborhood built in the early part of the twentieth century. In its early days, this was a lovely, upper middle class neighborhood. Set on a bluff 100 feet above the Hudson River, the area provided sweeping views across Fort Washington Park and many amenities. In the 1920's, both Columbia University Medical School and Presbyterian Hospital, two prestigious institutions, chose to develop their new, joint campus in this thriving area, becoming the centerpiece of the neighborhood.

Figure 9.1 New York Psychiatric Institute — top view. (Photography by Dan Cornish.)

By the 1990s, the Washington Heights area had developed a highly diverse population, mostly Latino, Black, and elderly Jewish residents. The area had become the center for immigrants from the Dominican Republic, by far the largest ethnic group. Economically, the area had declined, with one of the highest unemployment rates of any neighborhood in New York City, and with a high crime rate to match. Issues of greatest concern included the neighborhood economy, jobs, language issues, safety, park maintenance and access, as well as the availability of medical and psychiatric services for neighborhood residents. Vocal activists and political leaders including Assemblyman Denny Farrell, local leader and City Councilman Guillermo Linares, Community Board President Maria Luna, and others were sure to be concerned about the project's relationship to key community interests.

The Columbia-Presbyterian Medical Center had weathered this neighborhood decline, retaining its prominent reputation as a first class medical and research institution. Dr. Herbert Pardes, dean of the Medical School and former director of NYSPI, had been key in maintaining the institution's stature. However, the town-and-gown issues were in full force as the medical center had often attempted to expand into the surrounding residential territory. Although the medical center was clearly the largest economic driver of the area, residents did not always feel the benefits of this great institution. Anger toward the medical center for broken promises for local work participation on previous construction projects did not provide great credibility with neighborhood leaders.

The dilemma

As NYSPI looked for ways to expand, it became clear that there was absolutely no land available in close proximity to the institution that could fit their needs. An exhaustive feasibility study took place in the late 1980s in an attempt to renovate and expand in place. After several years of serious study, this was deemed impossible for cost, constructibility, and operational reasons. Again, the institute looked for land, but once again, no potential sites could be found within the neighborhood. As the search broadened, it came to the attention of NYSPI that two private parcels of land existed in Riverside Park and were owned by the medical school and the hospital, an earlier gift of the Harkness family. Each owned one parcel. No one had ever built a building west of Riverside Drive for the entire length of the park. It was seen as a continuous swath of parkland lining the western edge of Manhattan. With a size of 369,000 gross square foot, this was not going to be a small intervention. But perhaps this was the only option. OMH would need to make their case.

The NYSPI, through OMH, proceeded to make a swap with Columbia Medical School, essentially trading ownership of their present main building in exchange for the land across the street, one of the two privately held parcels. It may seem that as private land, a building could be built "as of right," as long as it complied with all zoning regulations. However, this was not the case. Several factors complicated the situation. In the 1970s, the land had been incorrectly "mapped" by the city as parkland, even though the city did not own these parcels. A mapped zone would require "demapping" by the legislature to be buildable – obviously a highly political proposition. At one time, the city had considered buying these parcels, but this purchase had never taken place. Nonetheless, it was clear that certain groups would maintain that it should remain as parkland; the environmental review process including the submission of an Environmental Impact Statement (EIS) would be the legal route for the opposition to derail the institute's project.

The strategy emerges

Early on, prior to 1991, the Office of Mental Health hired a special environmental and political advisor named Ethan Eldon. He was the key person in all matters of strategy, and OMH displayed vision to engage such a knowledgeable person as its first team member. It showed that they were serious about taking on this challenge. The team was then assembled, including the architect, the engineers, and the construction manager. The facilities development corporation, managing the project for the state, was active in supporting this overall coordinated effort guided by Eldon's strategic sense.

The process was very straightforward:

- Identify all interested parties.
- Determine the best sequence for meetings.
- Meet with each group or person to present the project, its purpose as well as its design.
- Listen to their issues and concerns, and address them, if possible.
- Try to gather as much support as possible.

Behind the scenes, another goal was to keep the project from becoming a pawn in the larger political landscape of city vs. state. In this respect, it had to remain a fairly low-profile matter. Political silence, if not support, became very important.

In presentations, there were no hidden agendas. To gain support, we needed to have credibility. The contacts and meetings had to take place in a very specific order, so as not to offend any particular party. Initially, Eldon met many parties privately, to find out how they felt about the idea of building a new building for NYSPI, west of Riverside Drive. By the time the architects were on board, there was a fairly clear picture of who might support the project, who probably would not, and who was on the fence.

In order for the project to succeed, we needed strong and committed internal support as well. Skeptics in Albany needed to be equally convinced that the project had a chance. The full support of Governor Cuomo and the OMH would be critical. In total, the list of interested parties was enormous. In addition to the Governor's office, and Dr. Richard Surles, the commissioner of OMH, it included Mayor Dinkins, the Borough President Ruth Messinger, numerous elected public officials, various New York City agencies, Presbyterian Hospital, Columbia Medical School, the local community board led by Maria Luna, various environmental groups, and more than ten local groups. Most notably, unelected but very important leaders from the community itself played key roles in rallying community support, including future City Councilman Guillermo Linares and Moises Perez.

Meetings ranged in location — from City Hall to private law offices on Park Avenue to the basement of local buildings. Meetings were sometimes conducted in Spanish. The experience was time-consuming, provocative, and influenced many peoples' lives and careers.

Plusses and minuses

Prior to the meetings, we tried to assess the situation. The Psychiatric Institute (PI) had some real plusses. Seen as a good institution providing needed services to the community, the PI wasn't quite viewed as a neighborhood intruder. About 85% of the employees were New York City residents; many came from the Washington Heights area. Services at the institute were provided in a multi-lingual setting, where both English and Spanish were spoken. PI was a public institution, and had tried to be "a good neighbor" over the years. Financially, the institute took in much more money in grants than it received from the state; therefore, it was a positive economic force for the neighborhood, the city, and the state. Finally, the highly accomplished reputation of the institute made it difficult to object to the need for expansion.

According to this logic, supporters should include advocates for mental health, other members of the medical center community in Washington Heights led by Dr. Herbert Pardes, the Economic Development Corporation in Manhattan (EDC), OMH, and the Governor. However, even these groups required convincing. In some cases, agreement was based on specific design issues, such as height limitations, maintaining views from surrounding buildings, providing an active street frontage, transparent building materials for a welcoming appearance, and massing to retain key natural site features.

Opposition was expected on other fronts — those who were not interested in supporting a controversial project, those who oppose any expansion of the medical center at all, and, most importantly, from environmental groups and park enthusiasts on every level. To many, taking down a tree — any tree — should be stopped. Past history has shown that the power of environmental groups is enormous.

The neighborhood itself was the biggest open question. Would they support or oppose the project? Opposition was assumed by most who were close to the scene. Neighborhood support was contingent upon the residents' view of the institute as either friendly or hostile; on PI's credibility to come through on promises for employment of local minorities in the construction process, and on the design itself — would it be an asset to the community? Initially, no one in the community differentiated between Columbia, Presbyterian Hospital or NYSPI. In the final assessment, it was the strong outpouring of neighborhood support, both for PI and for the design approach that made the project possible. This support was garnered by an inclusive and sincere effort to understand key design issues, and to mitigate those concerns through design, wherever possible.

One subtlety was that elite, downtown politicians, such as the borough president, were expected to oppose the project on the basis of the violation of perceived green space. However, in all presentations, we attempted to dislodge this notion of parkland through photos, models, and description. The site, although west of Riverside Drive, was poorly maintained, unsafe, surrounded by a spaghetti of ramps leading to the George Washington Bridge, and hardly qualified as a real amenity. It was "green" only when viewed from the air. In fact, it had been named "dead dog park" by the previous Parks Commissioner Henry Stern — hardly a complimentary term.

The design role

At the time the political and community presentations began, the architects were about 6 months into the project. We had developed a program of space needs for the building, and had prepared site analysis and a preliminary building concept — an idea of the overall massing of the project. The presentations consisted of a description and history of the institute, the services that were provided within, the state of the existing building and the need for a new building, a description of the site, and the proposed design concept. In all meetings, the impact on the community was discussed in an open dialogue. The team presented essentially the same material to all parties, in a travelling portfolio, ready to present on a moment's notice, day or night. The architects, along with leaders of the institute, took the lead in these sessions, under Eldon's tutelage.

The site and program

The site was located roughly 100 feet below the rest of the community, at the base of the cliff mentioned earlier (Figure 9.2). It was therefore disconnected from the bulk of the community, set adjacent to parkland to the north, south, and west. The famous views of the river and the bridge were not even possible from the site, as it was surrounded on three sides by ramps and highways. It was a very unpleasant place; included land that was created by landfill during the construction of the George Washington Bridge.

The site was very constrained for the program, which consisted of research beds, clinical beds, outpatient clinics, educational facilities, wet "bench labs" for basic science research, offices for human research, and even a public school for children in psychiatric research protocols. It was a complex program, and would have been an architectural challenge even without these site constraints. Architectural goals for the project included a low (six-story) profile, such that views to the river would be maintained for the existing buildings on the cliff above; a functional, state-of-the-art, efficient

Figure 9.2 New York Psychiatric Institute — corner of building. (Photography by Dan Cornish.)

building, and an expression that would give the institute a new image. In addition, the site was highly visible when entering the city via the George Washington Bridge. In effect, it became a "gateway building", impossible to miss in this highly visible spot.

The usual environmental issues, such as noise, traffic, shadows, and parking were not as critical in the political analysis, as the site was already isolated from the rest of the neighborhood. However, the surrounding roads, potential noise and vibration disturbances for research activities, and required parking and service access presented additional design challenges.

Not only was the site surrounded by highways on three sides, it was an odd configuration as well. The original scheme consisted of a six story structure plus a three story "occupied bridge" crossing Riverside Drive, connecting the new building to the existing Kolb Annex. Early in the meetings, it became clear that this design would not be acceptable to several key people. City planners did not like bridges, as they kept the people (and the city life) off the streets. An occupied bridge would be even worse, blocking views with a major building mass.

The presentations

The team's adopted strategy was to present the project clearly, listen to all players, and be respectful and inclusive of multiple views. The designers had to be flexible and knowledgeable of construction issues, codes, and program requirements. We needed to allow room for input into the design process, and be responsive to suggestions, not defensive of specific design ideas. The building expressed a clear architectural vision for the institute in an elegant, curved form facing the river, which seemed to excite the community. It was a new form, breaking from the rectilinear mass of masonry buildings in the rest of the medical center complex.

The presentations proceeded over many months. As the project progressed, the design was changed and refined in response to the various concerns. The most significant change required obtaining the second parcel of land from Presbyterian Hospital through the legal process of eminent domain through the New York state attorney general's office. This process effectively condemned the land and forced eviction of the owner's land, for use by a public entity for the larger public good. Presbyterian Hospital was reluctant but was able to get fair market value for the land in this arrangement, which was very appreciated. This enabled us to keep the design at six stories but eliminated the "occupied bridge." Instead, two smaller, pedestrian connecting bridges were designed. One connected both visitors and services to the Kolb Annex, creating a door to the community at the 168th Street level. The second bridge connected the institute to the Milstein Hospital building at the medical center, for direct patient and staff access. The Riverside Drive entry was widened to present a gracious drop-off, and the building set farther back from the street; the six-story height limitation prevailed throughout. The curved shape, facing Riverside Drive, was an integral part of the design from the very beginning. This never changed, as it represented the key signature design gesture for the new institute building. Uses were placed on the ground floor that would "activate" the street, creating a lively streetscape, with a transparent and dramatic atrium marking the entry. The presence of activity would enhance the general security in that section of Riverside Drive. And finally, the institute committed to pay for maintenance of a portion of the surrounding park, making it truly an amenity for local residents.

A positive response

At many moments, we were on pins and needles, awaiting the response of various parties, and for the EIS time limit for opposition to expire. Those whose support we counted on came through; those representing environmental groups were opposed to the bitter end. The borough president opposed the project, although not as vocally as she might have. Upon hearing the community's views, other public officials either supported the project or remained silent. The mayor's office expressed reservations, but never came out against the project. Assemblyman Denny Farrell, an influential member of the state legislature and head of the state finance committee, was particularly helpful and supportive, a powerful figure, savvy, intelligent, and very familiar with the site. He later embarked on a project to obtain funds to provide a connection to lower level park areas for Washington Heights residents during the process. But it was the community who came out in full force, eloquently defending the institute and its need for expansion. In both English and Spanish, the local residents came through in favor of the project. Their support went a long way to neutralize the opposition, which ultimately allowed the project to be built. The EIS was approved, and no lawsuits were filed. In many ways, what did not happen was more important than what did.

During the process, specific deals were brokered:

- The mayor's office, under Deputy Mayor for Planning and Development Barbara Fife, suggested that they would support the project if we eliminated one overhead bridge. It being an election year, however, the mayor declined to support the project outright, and the two bridges remained.
- Although the Presbyterian Hospital officials remained skeptical, the medical center supported the project based on the PI's commitment to a low building, thus keeping the medical center and its views.
- The community received a real commitment for construction jobs, enacted by establishing a storefront center to assist minority contractors in filing appropriate forms. The PI established an ongoing advisory committee to address community issues.
- OMH committed to trade outlying wetlands as ball-fields in other boroughs of the city, to appease the State Parks Department and other parks advocates.
- The south end of the property was retained as a public children's park for the community, with anticipated access to the river at 165th Street to be built with additional funding.

Epilogue

When I attended the opening ceremony in April 1998, I was seated next to an important and influential doctor who had been very involved in the design process. It was a festive atmosphere, and everyone was thrilled with the building. When asked about the new building and the state of the institute in general, he said, "It's too small — we've just moved in, and we've outgrown it already." While this is no doubt a great challenge and problem for the institute, it speaks to the great success that the project has had. The mission of the institute continues, and more and more grants are received every day, furthering psychiatric research developments, in the inspirational setting that they deserve.

The community has benefited as well. In its expansion, the NYSPI has continued its commitment to local employment. Today, the surrounding parkland is not the derelict, unusable "dead dog park" of the past, but is maintained and usable. Funding is still being pursued to build a pedestrian access through the improved park in order to reach the Hudson River, a concept that emerged through neighborhood discussions surrounding the NYSPI building project. And Riverside Drive, long an inhospitable vehicular route in this section, has been enhanced by the presence of the building itself.

Often I drive by the building on the "spaghetti" of ramps to and from the bridge. Although I am pleased with the building's dramatic form and appearance, I cannot see it without remembering the extraordinary process that preceded its construction. No one involved in the project seems to forget it — the building's existence almost comes as a surprise, even a small miracle. It is a true testament to visionary and committed leadership, to the importance of design, and to the value of a truthful, inclusive process.

chapter ten

Building under a city street

Deborah Leonard

Contents

Land use 108
Zoning 110
Environmental quality 111
Building codes 113

This chapter tells the story of an addition to the law school of a major urban university. The university is set in the middle of a congested area yet faces the overwhelming need for more space. After several years of deliberations by the school and calls for action by the university to satisfy its programmatic need, a team of experts were hired to tackle the issues and produce results. The team consisted of an architect, a construction manager, and legal counsel. Subconsultants to these primary members were added as the project progressed.

Do you have a lot of time, deep pockets, and a professional staff well-versed in working its way through agencies, divisions, departments, and commissions? Then you will be fine. If not, brace yourself and fortify your bank account before attempting the development of your privately-owned property, as it involves the world of the public domain.

To say that the necessary steps, procedures, and approvals are cumbersome and time-consuming is an understatement at best. Tenacity and perseverance are required — even for those who are seasoned in real estate development and know what's in store. And that's before you can even begin to imagine the construction issues and problems that will cause you to lose sleep.

Why would anyone consider pursuing a formidable course knowing the difficulties that will be involved? The hurdles are approached because they absolutely must be if developers, owners, institutions, and facilities are going

0-8493-7486-3/01/$0.00+$.50

to maintain their status within their field. Institutions or facilities — of all kinds — must maintain a level of service that places them at the top of the list, so to speak, and command high prices commensurate with meeting their expenses. A hospital, for example, now considered a profit-making business, must maintain its stature by offering the best, most-sophisticated, state-of-the-art treatment and facilities. In order to do this, adequate or additional space requirements are an unmistakable necessity.

In this chapter, we will take a look at an urban educational institution endeavoring to stay in the top echelon of the highly competitive world of law schools. In order to maintain its first-class stature, the institution must attract and keep premier faculty. Hand-in-hand with top-rate law faculty are a top-rate reference and literary collection, and library amenities. Student tuition is spent in support of faculty and other operating costs for the school. To give further credence to the costly tuition, the best facilities and services need to be offered.

In this case, the school was painfully aware of the need to expand their existing collection, and to provide space for historical volumes which were being kept in remote locations. The facility also needed to be updated and expanded to include space and infrastructure for the technological age: media rooms, computer facilities, and the accompanying electrical support. In addition to added square footage, the new space was to create an environment conducive to studying and reflecting. Great attention was given to the atmosphere that would be experienced by the users of the finished library– providing more natural light, augmenting natural light with strategically-designed lighting systems, and providing ease of access between the floors and areas of the library. These considerations are critically important in order for the final product to accurately demonstrate the premier stature of the institution and what it offers to its customers.

All avenues must be investigated for finding a way to expand. The cost-effectiveness and feasibility of the different design options required professional expertise and deliberate review by the development team. In this instance, the potential for building higher — on top of the existing facility — was not a practical option because the need to reinforce structural and foundation elements posed as great an undertaking as building "out" did. Also, the prospect of altering the scale of the surrounding neighborhood was not thought to be responsible from a planning standpoint. Positioning the new facility in a remote location was not the answer – there needed to be a facility "under one roof" for easy access to all data. Inasmuch as any expansion would involve a permitting process, it was sensible to pursue the option providing optimum results for the owner.

The option for the library expansion program that proved the most advantageous from both a design standpoint and from the perspective of future growth potential was chosen due to the existing circumstances and constraints, even though it had some drawbacks. The existing library was in a four-story building that was bounded by city streets in all four directions — north, south, east, and west. The library occupied three levels: street level and two levels below ground. It was seen that by expanding in an easterly direction, the owner would have additional benefits because they could then connect into one of their other properties located on the opposite side of the street. This scheme achieved the goal of providing library spaces to one another under "one-roof" — even if part of that "roof" was actually a city street.

The advantages and possibilities presented by this scheme were great enough to warrant the owner's pursuit of approvals required for building underneath a city-owned street, even though it increased the fees (legal and design) and the time frame. The owner was well-versed in local procedures and agency approvals, was no stranger to the local community and the objections likely to be raised against new development, and also knew how to address the objections and continue moving forward through the approval processes. The development team's experience with development and growth was extensive and the professionals engaged were able to navigate the bureaucratic quagmire. This was not a program to be undertaken by the inexperienced and the successful outcome of this project was due to the team's advanced knowledge of planning for potential pitfalls.

There were four major areas requiring approvals. They are outlined below, and explained in greater detail following. The categories are typical for the kinds of approvals cities require, although the agency names given here are for New York City.

1. Land use – Application to the Board of Estimate must be made in order to request permission to develop and use the area beneath a city-owned street. If accepted by the board of estimate, a franchise would be issued, granting the owner the right to use the property for a specified use, for a specified time, for a specified fee. There will probably be conditions that the owner must meet if the franchise is granted.
2. Zoning – To comply with local zoning requirements, the project must have an approval from the Board of Standards and Appeals, the organization responsible for amending or repealing rules and regulations pertinent to the building code and zoning regulations.
3. Environmental quality – The approval issued by the Board of Estimate will contain a condition that a city or state environmental quality review must be performed to determine the impact proposed construction might have on the surrounding area. The conclusions of this review will reveal whether or not additional actions by the owner are necessary due to any environmental impacts resulting from the proposed construction.
4. Building Codes – Standard construction approvals mandated by the department of buildings were to be obtained, and customary filings and approvals provided (e.g., building permits, department of highway permits for adjacent street and sidewalk storage areas, closing of sidewalks, approval of new sidewalk plans).

At first glance, it is evident that there are at least four agencies of specific authority from which approvals must be obtained. The timing for obtaining each of these approvals is critical. One objection is all it would take to derail the project. The owner had already incurred expenses, and begun some long-range planning and programming. What if the community objected in full force? The Board of Estimate only meets every other month and that approval is needed to go to the next step.

As in any city or community involving multiple approvals, an understanding of the processes proved a distinct advantage in this project. Had the owner been unaware of the intricacies of the different approval processes, he would not have known to pursue multiple approval paths, simultaneously, and the overall period for approvals would have been far more lengthy.

Being prepared for opposition to the development enables an owner to have reasonable explanations and options available for discussion. Some opposition may be well-founded and appropriate, but by planning ahead and investigating different options, an owner may be ready to provide quick alternates to a community or agency's objections. Being able to compromise may mean the difference between an immediate "positive" approval versus the time consuming process of going back to the drawing boards to react to opposition and then reconvening at a later date with the alternative.

Land use

In order to use the area beneath a city-owned roadbed, the owner had to comply with the directions of the city's building code: "Tunnels connecting buildings, and projecting beyond street lines, may be constructed subject to the approval of the Board of Estimate and the department of highways. Such tunnels shall comply with the provisions of this code and other applicable laws and regulation." Permission from the Board of Estimate is the first step in the journey through the approval process.

The development team applied to the Board of Estimate and subsequent agencies with a well-defined program. A clear description of the evaluations made by the development team was needed: what other design or expansion options were investigated; why did this particular one prove most efficient or viable; what will the end product look like; what improvements will it provide to the community or surrounding area. If this information is clear and complete, it is likely to answer any questions the reviewing agency may have about the proposal and thereby, controversy or opposition is reduced. The benefit to the owner is a quick approval. On our project, the exterior was most critical to the community and local agencies. They wanted to know: what will it look like when the street is restored; how much "better" is it; is public access provided; is landscaping given attention; and are there any community benefits? They were also concerned about what would happen during construction. How will traffic be dealt with during construction — after all, the street will be closed for many months during construction activities — how will this affect the community? Have traffic studies been done; what are the conclusions?

In our case, the obviously negative impact on the local community was the loss of a small "thoroughfare" during construction. This could not be denied. The best defense was to describe the duration for this imposition and keep to the schedule; updating the local community on schedule changes over the course of the project was also helpful. Using a proven rule of thumb that anything that adds to a "total picture" is helpful in describing the mission to the approving entity, the team prepared thorough descriptions. Some may think it is better to cloak the negative impacts, but from experience, the team recommended that both positive and potentially negative aspects of a project be set forth as non-embellished facts. The documents that present the development program generally become part of the final record of the approval because they define the scope of the authorized program.

Once the Board of Estimate has approved an application, a resolution is issued. The resolution from the board contains specific qualifications upon which the approval board is based. In this case, the resolution contained several conditions:

- The Board may discontinue the agreement if it determines discontinuation is necessary. The owner therefore allows the destiny of the building to be based upon the judgment of the board, a political body subject to change.
- The construction proposed would be for the sole use of the owner and only for the purposes described. Any future assigning of the space or subletting of the space must first be approved by the Board of Estimate.
- Should the owner decide to vacate either side of the "tunnel" the resolution is automatically terminated. This locks the owner into the program defined and approved in the resolution. In our case, the owner was able to agree to abide by this term because its immediate and long-term needs are exactly as defined in the resolution.

There are yearly franchise fees defined in the resolution payable by the owner to the city. A provision was included for adjusting future rates, and another directed the owner at its own expense to remove the "new" structure and restore the street upon demand by the board. The owner was willing to risk that this demand would never be made because of the dire need for space.

The resolution also states that the effects of construction on the surrounding area, including all existing services and utilities are the responsibility of the owner. In addition, all adjacent areas are to be protected from the construction and any alterations to existing structures for accommodating construction of the new facility are to be included as part of the owner's development program. The team extensively surveyed the existing conditions during the planning stage in order to apprise the owner of the potential costs. Protection of adjacent properties and infrastructure (e.g., shoring, bracing, support of utilities) were a major consideration, and costly.

Over and above the specifically stated conditions of the resolution, the owner must abide by the stipulation that all requirements of other appropriate agencies are in order. All agencies require the filing of applications, approvals, and permits before work can start and the cost and time for these can add up. Depending on the backlog of work at the individual agencies, a project may wait several months for approval of plans and a building permit allowing the start of construction. On our project, we accounted for these requirements up front, and identified their impact on the project schedule and costs.

While the resolution is in effect, the city requires indemnification by the owner for all costs resulting from claims due to property damage and personal injury. The university's need for expansion dictated the acceptance of this risk.

A franchise agreement or de-mapping the street are the only two vehicles allowing private use of a city street. The team did not propose the de-mapping of the street because of the anticipated unfavorable community reaction: the street was a busy route to a large park serving the entire community. But why would the owner enter into a franchise agreement with what appear to be one-sided and onerous terms? In this case, the owner and his expert consultants felt that the city could but would not introduce obstacles or encumbrances to the resolution. It could reasonably be expected that the city would not reverse the decision to allow the use of the street for the owner's expansion program.

As an interesting aside, the Board of Estimate was abolished approximately eight years after this resolution was granted. It was not replaced by the creation of another board or agency, but was added to the city planning commission's role and scope of responsibilities. Maintenance of the resolution became the purview of city planning, and has remained in effect.

Zoning

The design did not comply with the zoning requirements for the area, a fact that was known at the outset of the initial design phase. Under the city's charter, the agency responsible for interpreting the zoning resolution for the city is the building department. If a proposed project does not comply with the existing zoning requirements, the applicant (owner) must file an application to the Board of Standards and Appeals, the six-member board responsible for determining and varying the "application of the building zone resolution" (zoning resolution). In addition, applications for zoning variances are put before the local community board, and their opinion and vote is recorded and considered. Any objections are considered for validity and reasonableness. The variances requested in this case posed insignificant implications for the land lots involved: the existing zoning requirements required that the lots in questions be kept as open space, with no structures or obstruction. The design of the new underground facility required an emergency exit stair bulkhead at the back corner of one of the lots, so a variance was required for the placement of a small masonry stair enclosure in "open space" territory. The team sought a zoning variance while the application for the franchise was before the Board of Estimate. The owner's application for the zoning variance actually stated that the owner was simultaneously filing a separate application to the Board of Estimate for a recoverable consent to construct a facility under the bed of the city-owned street. This tactic was a significant time-saver for the owner. This zoning review — review by the board, input from the local community board, board meeting postponements, etc. — took five months, but coincided with the period for the Board of Estimate review. Some have been known to go on for more than a year. This is a five-month wait for an approval of insignificant impact to existing regulations. Nevertheless, the process prevails.

The owner has now received the two major approvals that allow the development to become a reality, and there is some assurance that the project can move forward. The next round of approval processes are those referenced in the resolution granted by the Board of Estimate where it was stated that "other applicable laws and regulations" were to be sought and obtained by the developer.

Environmental quality

One of the agencies that will be involved is the Department of Environmental Protection (DEP), which will perform an environmental quality review to determine what environmental impact the development may have on the surrounding area. The DEP will issue its findings in the form of a declaration, and in this case, a "conditional negative declaration" was issued: "negative" meaning the impacts were not problematic and "conditional" meaning certain conditions were attached to the approval. The two conditions in the negative declaration were: a) that the owner must give "concern" to vibration, water table, and archaeology, and b) noise abatement program was to be instituted and maintained during construction activities. The basis for requiring an archaeological investigation was the proximity of a historic landmark building and the noise abatement requirements were necessary because of the residential neighborhood surrounding the development city. Luckily, traffic and parking did not surface as major concerns of the community.

Concern regarding the vibration, water table, and archaeological issues was further defined in attachments to the negative declaration, identifying three specific areas: 1) concern for pre-construction archaeological conditions, 2) concern for pre-construction conditions in general including the existence of a landmark building, and 3) attention in monitoring the steps taken for theses two items throughout the construction program. To comply with the mandate for attention, concern, and monitoring related to archaeological conditions, the owner was obliged to perform a study and submit a written report of the historical significance of the site and its potential archaeological significance. Depending on the size of the site, the costs can be in the hundreds of thousands of dollars for the tools, staff, special equipment, and support personnel to operate machinery. The report included a determination of whether an archaeological field investigation was warranted. The report was submitted to the Landmarks Preservation Commission (LPC) (yes, another agency) which provided a final judgment on whether a field investigation was required by the owner, at the owner's expense. A field investigation was found necessary by the LPC, so the owner prepared a plan on how the field investigation would be conducted and submitted it to the LPC.

So, the costs are mounting as time is expended in obtained approvals before anyone puts a shovel in the ground. Approvals are being sought from and reviewed by groups that either have no interest in whether or not development goes ahead, or even have very strong positions countering the proposed development. It is important to know as much as possible about the opinions being voiced and what specifics are behind those opinions. Addressing one small issue or making one small compromise is sometimes all it takes to overcome flat-out opposition to the overall program.

At this point, a firm was hired to develop a field investigation program for submission and approval by LPC. The program defined the steps for the excavation and recovery process. It included a description of the laboratory set-up for investigating any findings and a schedule outlining the time for the field investigation. When the field investigation was completed, the archaeological team had one year within which to analyze their findings and submit a report. This means the owner must wait another year after the field investigation while they await the approval of LPC. Fortunately, the dialogue during the investigation process between the investigation team and the LPC — people who know one another professionally — enabled the LPC to sign-off the program and its results long before the field team's final report was submitted to them.

Archaeological investigations do not move expeditiously given the painstaking measures for careful digging and recovery of items. Incorporating the time for the archaeological field investigation into the overall project schedule is difficult because of the lack of a standard protocol for field investigations. No significant recoveries or surprising discoveries were made as a result of this particular archaeological mitigation program - yet it still moved at a snail's pace with no seeming schedule accountability on the part of the LPC. If there is a significant archaeological find, buildings may need to be redesigned around the find to leave it intact or the project may be stalled while removal of artifacts is done. Foreseeing the total impact from such an issue is impossible, no matter how experienced a team may be. This type of exploration or investigation has occurred more frequently since the early 1980's when notable discoveries were made such as the discovery of a boat unearthed during building excavation at the southern tip of Manhattan. Prior to that, there were few formal programs for assuring that findings were recorded or even made available to local historians for evaluation.

Although archaeological investigation seems just one more hurdle for land developers to jump over, the significance of recovered historical information is assuming greater visibility. Public support is increasing for preservation of archaeological sites and developer's activities may well be curtailed further. Archaeological explorations are more common and field investigations are regularly recommended and performed. It is crucial to be prepared for such possibilities.

Building codes

Before construction may start, permits must be secured for specific construction activities, most of which are dictated by the building code. Design drawings for the proposed project must be filed for consideration and approval by the commissioner of the building department. For this building, a complete set of drawings - architectural, structural, mechanical, electrical, plumbing, and fire protection needed to be filed. Plans are examined by the department under the supervision of the commissioner and when they are accepted, they are stamped approved with the official seal of the department.

In this particular case — as in many projects — the foundation design was completed well before the other disciplines. Architectural and mechanical, electrical and plumbing designs continued to be adjusted and refined as the program for the building developed. An application was filed for a separate excavation/foundation permit, allowing for the start of construction while the design was being finalized for the remainder of the building. This was a significant timesaving procedure.

In order to file an application for securing an excavation/foundation permit, many documents must accompany the application:

- Lot diagram showing zoning compliance.
- Complete foundation plans showing the size, height and location of the foundation, dimension of the foundation in relation to lot lines and streets, existing curb elevations and final grade elevations of the site when work is complete.
- Boundary survey prepared by a licensed land surveyor.
- Indication of city planning commission approval (and any other pertinent approvals for a particular site, as in this case, the approval of the Board of Estimate franchise agreement for the use of the city-owned street).
- Soil boring logs – certified by the owner's engineer that the borings were performed under his inspection and the data submitted is accurate.
- Design for underpinning existing, adjacent structures by the owner's engineer and inspection throughout the underpinning process by that engineer who attests to the accurate performance of the work according to the design.

- Soil bearing capacity before placement of concrete footings, walls, and piers-inspected and accepted by the owner's engineer.
- Concrete design mixes according to the strength of concrete required.
- Certificates of inspection at the concrete plant signed by the owner's engineer.
- Certificates of inspection at the site for placement of concrete and reinforcing steel signed by the owner's engineer.

This data is recorded indicating use of proper materials in appropriate quantities, and proper handling and placement of the materials in accordance with accepted standards for those materials. These inspections and certifications are not to be taken lightly, because they can protect the owner and designer if any improprieties are discovered in the future.

Several individuals actually perform the functions noted above as by the "owner's engineer." Usually, the inspection of the reinforcing steel is performed by the structural design engineer responsible for the entire project (engineer of record), and most other inspections for concrete are turned over to a controlled inspections firm's engineer selected by the owner, and acceptable to the architect and engineer. The inspection reports for concrete mixing at the concrete supplier's plant, concrete strength tests and on-site concrete placement are furnished by yet another engineer who is hired by the design engineer.

The costs associated with these approval processes are significant - involving design and consulting fees for an architect and other design team members; legal experts, code expediting experts, zoning counsel, time for the owner's representatives to appear before boards and community groups, costs for archaeological experts and their staff working in the field, and financing costs for the project. The percentage of such costs when viewed against the entire costs of a project vary, but generally run between 10 and 20% of the construction cost.

After the above approvals were received, the project encountered an unexpected set-back. An existing structure shifted at one corner of the new construction site, as revealed when residents of the existing five-story structure noticed cracks on their interior walls. Upon investigation by the construction team and monitoring consultants (those same consultants that were required by the Conditional Negative Declaration issued as a result of the Environmental Quality Review), it was determined that the interior cracks were a result of building movement. As a precautionary measure, steps were take to reinforce the exterior wall in question before proceeding with underpinning. All the steps proved successful and no further movements were detected.

The existing building that experienced the movement was in fact the property of the same owner of the new development. Nevertheless, the reaction to the situation was extremely cautious - all the occupants of the building were evacuated except for the resident building manager. If the situation was so dangerous, why was anyone allowed to stay? Or, was the evacuation forced because of the potential exposure to insurance or legal claims, if, by chance, something were to happen? Insurance is another arena affecting construction and the possibilities that arise throughout the life of a construction project. Is constructing really only a minor part of the development and construction process? At times it seems so; you set out to develop and construct, become engulfed in the processes and legal issues, insurance issues, financing issues and before you know it, construction seems like an afterthought. The costs for construction are customarily regarded as exorbitantly high — yet if one considers the costs for insurance to protect against a litigious environment, the financing costs for borrowing money, the taxes that will be paid over time, the costs to the legal experts to protect each entity from the other, then perhaps the cost for actual building material and labor are far less onerous.

People seem to remember the "disastrous" events given public media attention for construction activities, and when there is the merest hint of the potential for any such possible incident, the immediate reaction is to do anything to avoid danger or claims or loss by adjacent properties or the public. In the example recounted here, perhaps the evacuation of the building in question was not necessary — no further movements were ever detected and work proceeded without any problem. However, in the current political and administrative climate, the credo that it is better to be safe than sorry is the safest for pertinent officials to adopt. The evacuation had significant costs attached. Relocation for temporary housing costs for many people - costs for which the contractor's insurance carrier is responsible if the contractor is indeed found culpable, added hundreds of thousand of dollars. Interior repair costs were relatively minor, consisting of interior plastering and painting.

At the outset of any such situation, it is important to consider how the costs for such delay can be assessed; who handles the public relations, if necessary; will the nature of the situation affect the ability to move forward by demanding so much attention that it takes away from the normal workings on the project? These factors may be even more significant than the "damage" or "emergency" itself. The rest of this job was not allowed to come to a halt because corrective measures were taken promptly and excavation work continued immediately after. Taking well-considered control of such an occurrence by acting quickly, aggressively, and positively is the best response.

There are many instances of aggressive, irresponsible, and arrogant development and construction in cities. While the locale and particular project may change, the issues that need to be examined bear remarkable similarity. For example: consideration to the existing scale of a community and larger or higher buildings among low-rise residential structures; consideration to the surrounding community and the community's enjoyment of open areas and natural resources; consideration about the construction process - can work occur at all hours of the day and night? Can heavy trucks roll in and out of an area blocking sidewalks and streets without regard to pedestrians and other traffic? Regulations and conditions effect a compromise between a project and its neighbors. Development teams do not often reside and work in the neighborhoods which they are developing. Most likely they do not, making it even more critical that there be representation of those who have a vested interest in the area so that there is a level of comfort in knowing that attention is given to the quality of life in communities. In this case, the owner's reward for all of the conditions and costs expended was a dramatically improved facility attracting both staff and applicants for years to come.

chapter eleven

We've got an historic landmark, now what do we do?

Richard W. Southwick

Contents

The historic building design process 119
Historic research 119
Survey and assessment 120
Landmark approvals 120
The historic building construction process 121
Construction procurement 122
Construction team 122
Site logistics 123
Partnering 124
Case study I 124
Case study II 128

> "When we build let us think that we build forever. Let it not be for the present delight, nor for present use alone; let it be such work as our descendants will thank us for, and let us think, as we lay stone on stone, that a time is come when those stones will be held sacred because our hands have touched them and that men will say as they look upon the labor and wrought substance of them, see! This our fathers did for us." (John Ruskin, c. 1850).

0-8493-7486-3/01/$0.00+$.50

The art of building is timeless. From the earliest civilizations, buildings have provided the link of cultural continuity between generations, a concrete window into the societal structure and aspirations of times long gone. Ruskin's words, written a century and a half ago, understand the cultural lessons that historic buildings can teach, and convey the importance of sustaining older structures to be able to provide these lessons for future generations.

The majority of the older, and historic, buildings in the United States are found in urban centers. From the older dense central business districts in the cities of the northeast and midwest, to the historic cores of cities of the west and new south, these cities provide the link to the historic past for their communities. The familiarity of the traditional forms of historic buildings are a recognizable and comforting environment for users and visitors alike. Saving historic buildings leads to the greater retention of original urban fabric, maintaining streetwalls, similar building heights and massing, and an overall neighborhood coherency.

Historic preservation means more than saving individual historic buildings. Indeed, celebrated landmarks need to be recognized, protected, and restored. What would Chicago be without its Water Tower, Philadelphia its Independence Hall, or Atlanta its City Hall? However, the retention and rehabilitation of whole districts and neighborhoods are what give cities their distinct character and quality. This chapter will explore issues related to the reconstruction of historic structures within the urban context.

Historic preservation makes sense. It provides new uses for older buildings where public infrastructure already exists rather than building in new, undeveloped rural and suburban areas. This infrastructure includes mass transit, electrical, water, and sewer facilities, and the police, fire, and educational operations to service them. It provides housing and business activities such as manufacturing office and retail where people already live and work.

Historic preservation projects provide more jobs, and more jobs locally, than new construction. Typically, new construction costs are 50% materials and 50% labor. The production of sheetrock in Texas or windows in Iowa doesn't help the local economy. A typical rehabilitation project is more labor intensive and might spend 70% of its budget on labor, as provided by the local carpenters, plumbers, electricians, and laborers. This money in turn is reinvested back into the community.

Preservation projects are the original "green" architecture. Reusing buildings rather than demolishing and building anew is the ultimate act of recycling. One quarter of landfill sites are made up of construction debris; much of this from building demolition. Recycling structures significantly reduces this. Even a complete building "gut" rehabilitation saves much of the structural system, floor slabs and facades. Less intensive renovations can reuse interior partitions, stairs, and shafts. Oftentimes, historic and hard to obtain materials such as brick, wood planking, and ornamental metalwork can be salvaged and reinstalled. All of this reduces the construction debris generated by a building project.

Traditional buildings are constructed according to time-honored "common sense" design principles. Operable windows, overhangs for solar shading, and high ceilings providing natural convection are common in many older structures. Coupled with contemporary insulated wall assemblies and window construction, and new heating and air conditioning installations, rehabilitated older buildings can be energy efficient and environmentally sensitive structures. The taller floor to floor heights, common in older buildings allow for the introduction of new mechanical and electrical systems within either raised floors or lowered ceiling installations. The large shafts utilized for primitive ventilation systems of buildings one hundred years ago can be adapted for new HVAC and electrical chases or even elevators to meet ADA accessibility requirements. Overall, the older building is remarkably adaptable for contemporary use.

The historic building design process

The process of developing the design and contract documents for a rehabilitation project is very different than for a new construction project. The architectural and engineering teams must be competent in a whole array of additional specialized skills related to working on historic buildings.

The skills range from surveying and assessing the existing conditions of the historic structure, to researching and understanding the building, its provenance and the historic context in which it was built; to being able to apply older building and fire rating codes to its design and details. The ability to make convincing presentations and undertake negotiations with landmarks agencies is critical. One must be able to work with and adapt historic materials as well as specify and source appropriate replacements. The method of generating construction documents differs from new construction. Dimensioning and quantification of scope must be conveyed to reference the existing historic fabric being retained. And lastly, the architectural and engineering teams must plan on spending more time in the construction administration phase than the 20% recommended by the standard AIA contract, to respond to clarifications and unforeseen field conditions.

Historic research

Understanding the historic structure to be rehabilitated or restored is essential. Collections of original architectural drawings can often be found at sources such as architectural libraries, building department records, or corporate and industrial archives. Historical photographs and original documents such as correspondence, invoices, and newspapers, and periodicals can be located in local libraries, museums, and private collections. All of these are critical elements in piecing together the historic and physical components needed in a restoration project. Entrances and light fixtures, often elements removed from buildings in modernization projects, can be reconstructed from photographs and drawings. Additionally, original building components are sometimes, though rarely, obtained at architectural salvage yards. Learning about the building's era and designers allows the contemporary architect to make more informed judgments in his design decisions. For example, if the building under restoration is similar to more intact structures by the same architect, clues to the planning concepts and decorative detailing can be culled from the related building.

Survey and assessment

Equally important to understanding the background of the building is having a thorough knowledge of its current condition. Teams must obtain or reproduce existing condition plans, sections, and elevations, verifying drawn dimensions with actual field measurements. This painstaking and time consuming work includes documenting large scale elements such as partitions and openings to detailed items as wood trim and window profiles. This work can be facilitated utilizing new technologies such as photometric translations of images into CADD drawings, and low-intensity three dimensional laser imaging devices which can model large scale and hard to access interior spaces.

Traditional investigative probes are generally required to assess the condition of a building. Examples include removing flooring or roofing to examine slabs, opening up column enclosures to assess and measure structural steel components, and cutting open ornamental ceilings to review concealed mechanical and structural systems. Here again, more modern technology can be used instead of the traditional sledgehammer and prybar. Many of these new methodologies minimally affect the historic fabric of the building being examined, and are categorized as non-destructive diagnostic testing. Several more commonly used techniques include the boroscope, a fiberoptic viewing tube which can be inserted in a one quarter–inch hole into a ceiling cavity and manipulated to view concealed structural and piping conditions, and x-ray and ultrasonic soundings which can determine the extent and condition of hidden steel structural members without dismantling the enclosing masonry construction. Infrared photography can detect façade "hotspots" or heat leaks which can indicate a deterioration or failure of a building material behind the elevation. Hydrometric and spectrographic analyses can be used to identify moisture content and material degradation.

Many of these applications require expensive equipment, usually rented, and sophisticated interpretation, yet can still be considered economically feasible when weighed against the cost of probe dismantling and reconstruction, and the opportunity to provide a greater degree of certainty in the restoration process.

Landmark approvals

Many older buildings are located within historic districts, or if significant enough, are individually designated historic landmarks. The obtaining of the required approvals, often a Certificate of Appropriateness, can add considerably to the overall permit review schedule. The application process must be strategized and well-managed.

Cities, unlike rural or suburban areas, usually have local landmark preservation or historic review commissions. In addition, if state or federal funding is anticipated or an environmental assessment is undertaken and the property is listed on either the National Register of Historic Places or its state equivalent, the State Historic Preservation Office (SHPO), these organizations must also approve of the building plans. An application for the historic property investment tax credit program is also administered by the SHPO and will require their participation and approval.

The state review follows the proscribed Secretary of the Interiors Standards for Rehabilitation, commonly known as a Section 106 review, and applies a relatively strict set of guidelines dealing with both the interior and exterior of the building.

Alternatively, local municipal reviews often evaluate only the exterior treatment to the building: alterations, restoration techniques, demolition, and additions. The current trend on the local level is to draft concise and clearly understood design criteria against which to judge historic building applications. This helps guide the applicant through a more predictable approval environment than one based on more subjective opinions of its commissioners. Acceptable palettes of materials and color, and rules for items such as awnings, signs and air conditioners are established especially suited for the characteristics of the specific historic district.

Occasionally, landmark approval on both the state and local levels are required. This complicates the process as the separate agencies may have different preservation priorities. Rarely and usually only on very large and significant projects, the SHPO and the local landmarks agency will combine their efforts in conducting a single rather than parallel series of public hearings and coordinate review comments. Further coordination may be required if the appropriate design solution contradicts current, and usually non-contextual or traditional, zoning requirements. For example, if a streetwall height or setback rule requires an addition to be set on top of an existing rooftop cornice rather than behind it as requested by landmarks, a zoning variance would be necessary to implement the landmark approval. The further administrative reviews all can add considerable time and expense to the overall project schedule. Making changes to the architectural plans to conform to approval comments is often out of phase, that is, during the last parts of construction documents, and can result in delays in the construction issue date. In summary, the timeframe for the landmark reviews must be realistically evaluated and the owner and designers must be flexible and willing to negotiate with the historic agencies in obtaining the required approvals.

The historic building construction process

The construction process involving a historic property is considerably different than that which is required for new construction. The historic process can entail any of a number of types of intervention:

restoration – a historically authentic repair or reconstruction.
rehabilitation – a reconstruction retaining a varying degree of historic material while upgrading the building for current use.
dismantling and demolition – the removal of parts of a historic property, either carefully removed from a portion of the building to remain (dismantling), or the complete removal of a part of the structure (demolition).
addition – the new construction of a part of the project appended to the historic building.

In any of these construction types, a higher level of partnering is often required between the architects and engineers and the builder than in new construction. The full project team should strategize a construction buyout methodology specific to an historic presentation project. Specialty subcontractors and craftsman need to be identified and incorporated into the overall building team. And lastly, the tight urban site logistics must be planned on a building site not vacant but further complicated with a structure already in place.

Construction procurement

Bidding and buyouts on a historic project are generally more difficult because the extent of the work is not fully known until the construction is in progress. The exact quantities of many of the restoration trades can only be estimated during the preconstruction phases. Probes, hands-on investigations, and extrapolations of these findings are used to assign quantities for trades such as masonry repairs, replacements, repointing, and cleaning; structural steel replacements, reinforcement and repairs; and window and entrance restoration. The more extensive the pre-construction probes and testing, the greater the ability to accurately predict the quantity of restoration work required on a project. Full façade examinations will uncover many of the masonry repairs, and help to estimate the percentage of repointing needed on the facades. Test panels can fine tune the specifications of the cleaning procedures, identifying the most appropriate materials, proper dwell times, and number of applications. This information, developed from the on-site investigations, can be used to establish a baseline quantity of work. This can subsequently be part of a lump sum contract price. Any quantity of work in excess of or less than the contract amount can be adjusted based on pre-negotiated unit prices. This is generally a more cost–effective method of procurement rather than buying out this work on a net cost (time and material) or solely unit price basis.

Construction team

In addition to the typical subcontractors found on a new construction project such as the steel and concrete subs, the mechanical, electrical, and plumbing trades and the finish contractors — carpenters and painters, there is a whole host of specialty contractors required for a preservation project.

The masonry restoration subcontractor is oftentimes the key member of a successful restoration team. Specialized tasks such as composite patching, dutchman repairs, terra–cotta restoration, replacement, and façade cleaning require highly skilled workers and well-planned sequencing of the work. Replacement of stone panels and terra–cotta units necessitate an early identification and quick shop drawing and sample approval due to long lead times.

Similarly, window restorations are also very time-consuming and require craftsmen of considerable experience to execute properly. Specialized plaster contractors and painters are needed to restore interior decorative finishes more sophisticated than in general construction. A work force is often assembled not accustomed to working on a larger urban construction project and includes artisans such as stained glass artists, ornamental metalwork craftsmen, wood and stone carvers, and trompe l'oeil artists, among others. Usually, these craftsman are not union members and special considerations need to be extended by the union jurisdictions overseeing the project. Work hours are often longer and later than standard construction times and accommodations need to be made on–site for these special work conditions.

The integration of these artisans and craftsman into the pace and vision of the overall project is essential. The experience and sensibility of the site superintendent is particularly critical as the position takes on the additional roles of interpreter and educator for the less experienced trades, both the specialists and the general subcontractors.

Site logistics

Building on an urban site is difficult. Building on a site which already has an existing structure is even more challenging. The site logistics on an urban, historic preservation location are typically quite constrained and complicated. Older buildings often have a higher percentage of lot coverage than more contemporary structures. This greatly limits staging and delivery areas for storage of equipment, materials and contractor trailers. These functions are usually located within the building, blocking out large areas of the building which cannot be built out in the initial phases of construction. The cost and disruption of relocating these staging areas towards the end of construction is an additional burden on the project. This is required to complete the buildout in the original staging and delivery areas of the building.

It is very common to perform construction on an older building while it is at least partially occupied. Working around existing occupants necessitates finding temporary swing spaces, constructing protective barricades and enclosures, and providing temporary services while new permanent systems are being installed. This results in projects which are longer, more costly, and multi-phased.

Partnering

It is essential in the historic preservation project that the owner, the builder, and the architect/engineer act as a team, supportive of each others' efforts and understanding of each parties' priorities and perspectives. The nature of this type of project requires the early participation and cooperation of the contractor with the A/E team early on in the project to assist in probes and investigations, and to help develop buyout and logistics strategies before the construction phase. Engaging construction managers for preconstruction services allows for this type of assistance.

During construction, expeditious decision making is required as unforeseen field conditions arise, so as not to delay the course of construction. The builder identifies non-conforming conditions and must work quickly with the architect or engineer to develop and cost possible solutions for owner approval. The input of each member of the team with their own individual experience and perspective is valuable in generating workable design resolutions. The A/E team should expect to spend a considerably greater amount of time in the field than during a new construction project. A biweekly project meeting and weekly site visit generally will not suffice for a project entailing major building alterations, masonry restoration, and intricate interior finish work. Large scale historic preservation projects often require a full-time staff to attend to the field situations, observation, and restoration reviews which occur continuously during the construction phase. The design team's role becomes much more "hands on" as they work side by side with the builders on the project.

Case study I
Merchandise Mart Retail Development
Chicago, Illinois

The Merchandise Mart retail redevelopment project is notable for both the magnitude of the undertaking and the remarkable fact that the entire construction was completed in less than a year in a fully occupied building. The Merchandise Mart is located on the Chicago River, just north of the Loop on a prominent site in Chicago's River North neighborhood (Figure 11.1). Originally constructed in 1931, designed by the architectural firm Graham Anderson Probst & White, the building totals 4,200,000 square feet in area and is the largest single commercial building in the United States. Any design or restoration decision, and its attendant cost, is multiplied many times over on a building of this size.

The retail redevelopment project involved the conversion of 410,000 square feet of space on the first and second floors of the building into the shops at the mart retail center. The key to this conversion was the removal of 55 loading docks located on the first floor (upper deck level in Chicago's Loop) to the street level below, creating the opportunity of developing new

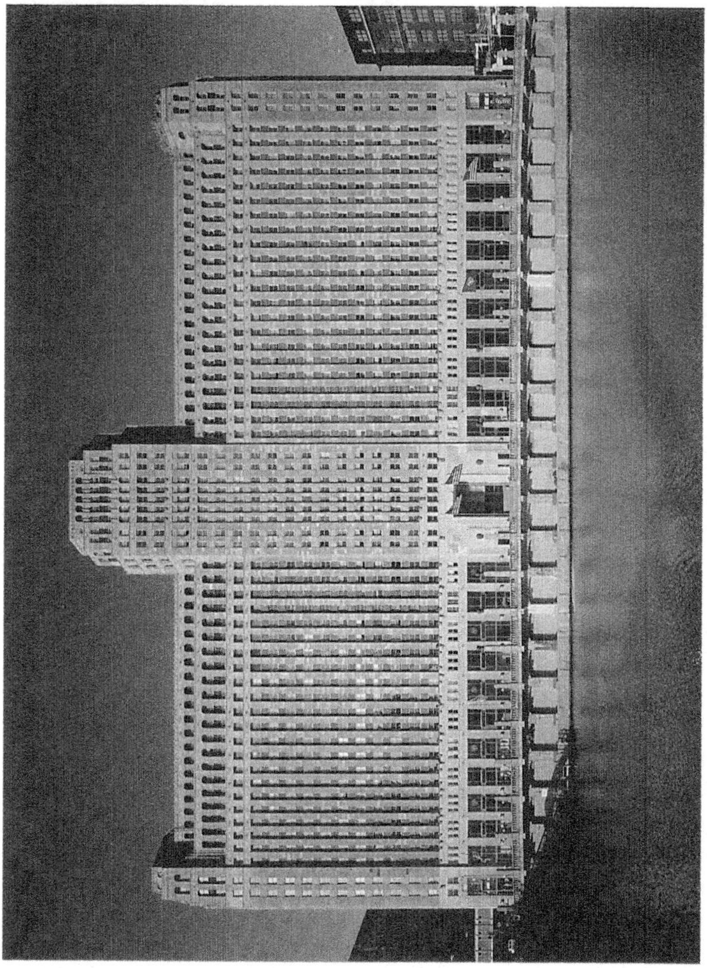

Figure 11.1 The Merchandise Mart, located on the Chicago River.

Figure 11.2 The existing loading dock area before the project began.

retail space on the north side of the building (Figure 11.2). At the time of construction, the River North neighborhood was emerging as a new "Soho-like" residential-art gallery district. The south side of this enormous building faced the central business district "Loop." New entrances and a large multi-height space in the original loading dock area opened the building up to this neighborhood and announced the transformation of the mart from a closed "to the trade only" building into a more mixed use, public structure. Coupled with newly restored and transparent storefronts, and entrances at each corner of the 800 foot long building, the mart now serves the surrounding neighborhoods as well as becoming the world's largest design and contract furnishings showroom center.

The Merchandise Mart's extensive historic archive of original drawings and photographs were immensely valuable in reconstructing the original entrance, storefronts, and sculpted limestone archway. Portions of photographs were enlarged to help detail the ornamental carvings above the main entry. Beyer Blinder Belle's design approach was to restore existing historic finishes and reconstruct missing elements in the original, southern part of the first floor lobby arcade and storefronts. The new portions of the project, such as the first floor north lobby and arcade, and the second floor, have a more contemporary design, interpretively derived from the Chicago School art deco character of the building, yet specifically distinguishable from the restored areas of the mart. For example, the palette for the new terrazzo flooring contains cleaner, more refined tones than the original muted color

Figure 11.3 The new retail center in the former Merchandise Mart loading dock area. Note the truck dock height on the upper level.

scheme. The new common area light sconces are a bronze–toned aluminum in a more modern, spare design than the original ornate bronze fixtures (Figure 11.3).

Constructing the retail center on the lobby entrance floor while fully occupied was particularly challenging. Eight major construction phases were planned, with logistical plans that constructed one half of the common area corridors at a time. In addition, more than 250,000 square feet of tenant improvements for 85 stores and restaurants were constructed at the same time, and had to be coordinated with the base building work. The $50 million project was part of an overall building rehabilitation that also included stone repair and replacement, new windows, and upgraded building mechanical and electrical systems (Figure 11.4).

Project Team:

Owner:	The Merchandise Mart Properties, Inc.
Design Architect:	Beyer Blinder Belle
Production Architect:	Jack Train Associates
General Contractor:	Pepper Construction Co.

Figure 11.4 The new north lobby space at the Merchandise Mart, with an entrance out to the residential River North neighborhood.

Case study II
Henri Bendel Retail Store
712 Fifth Avenue
New York, New York

The construction of the flagship Henri Bendel store at 712 Fifth Avenue in Manhattan was a challenging undertaking not only due to its congested midtown location but also because of its complicated maze of required approvals. The project entailed the buildout of 80,000 square feet of high-end retail space within the base of a new 52 story concrete office tower. The first 50 feet of the project were located in three five-story traditional Fifth Avenue townhouses, with the tower rising behind and above the landmark structures (Figure 11.5). Originally slated for demolition, the historic structures were identified and saved during the middle of the design process.

The three Fifth Avenue townhouses (712, 714, and 716) are quite different from each other. 712 Fifth Avenue is a limestone-faced cast iron framed structure from 1908. Its immediate neighbor to the north, 714 Fifth Avenue, has a steel and white marble façade in front of a structure consisting of large, load-bearing brick masonry piers and heavy timber floor and roof framing.

Figure 11.5 Three Fifth Avenue townhouses with the office tower behind. Note the new building (716 Fifth Avenue) at right of photo.

A three-story high, cast glass ornamental window by the renowned Parisian artist René Lalique is the 1908 building's most distinguishing feature. Obscured and forgotten for many years, the discovery of this unique and invaluable artwork prompted the New York City Landmarks Preservation Commission to designate 712 and 714 Fifth Avenue as historic landmarks. The remaining site at 716 Fifth Avenue was an unremarkable two story 1960s retail shop. This was demolished for the construction of a new five-story townhouse to complete the Fifth Avenue ensemble. The new structure is a limestone infill building, traditionally designed to relate to the similar 712 Fifth Avenue. It acts as the other solid "bookend" to the highly transparent Lalique window at 714 Fifth Avenue, located between the 712 and 716 structures. Upon closer examination, although matching 712 Fifth Avenue closely in material, scale and rhythm, 716's detailing is quite spare and contemporary. For example, fully three-dimensional balustrades at 712 are rendered more abstractly as two-dimensional forms at 716 Fifth Avenue. A similar approach is evident in the storefront, balconette, and dormer designs.

The reconstruction of the two landmarks at 712 and 714 Fifth Avenue was undertaken with great care. The difficulties of modifying an archaic cast iron frame with terra–cotta arch floors required investigations into the most appropriate methods to preserve the early twentieth century construction. At 714 Fifth Avenue, the wood framing was removed due to fire safety requirements. The removal of the framing plus the shearing back of the masonry piers allowed the construction of a central four story atrium which unifies the three townhouse structures and the larger part of the store located to the rear of the landmarks. While parts of the piers were being removed, and prior to the reattachment of the façade to the new permanent structure, the landmark elevation including the Lalique window was braced back to the tower framing fifty feet into the store. Survey measurements of the façade position were taken twice daily during this critical period to verify that the building was stable and had not shifted.

The project approvals were equally complex. New York City Landmarks approvals were required for all exterior and interior work within 50 feet of Fifth Avenue, that is, within the portions of the original buildings being retained. The interior design of the store was all new. In this case, it was unusual for an entirely new interior space to fall under the purview of the landmarks agency. Landmarks was also very involved in monitoring the restoration of the Lalique window. Consisting of over 200 lights of 15% lead crystal cast glass in a shallow bas-relief, a large number of glass panels required replacement or repair.

The New York City Planning Commission was also very involved. In this highly congested part of the central business district, the city required a mandated passageway through the first floor of the store to mitigate the increased load of pedestrian traffic that the new fifty-two story tower would generate. Because the sidewalks could not be widened due to the retention of the landmarks at the property line, an alternate path was established through the building, open to the public during the store's normal operating hours.

Figure 11.6 Four story Henri Bendel atrium with the Lalique windows overlooking Fifth Avenue in the background.

Finally, approvals from the Building Department were necessary for the smoke evacuation and makeup air requirement for the 714 Fifth Avenue atrium (Figures 11.6 and 11.7). This four story open space required water

Figure 11.7 Entrance doors at 714 Fifth Avenue leading to the Henri Bendel atrium beyond.

curtains and six air changes per minute during a fire emergency. Normally, large fans and louvers would be incorporated into a new building exterior wall. This was not possible within the landmark façade. Instead, a series of transom windows were modified to open electronically just above street level to provide makeup air. A series of hatches and spring operated skylight bulkheads at the top of the atrium allows the smoke to escape the space. This ingenious approach preserved the landmark façade while meeting stringent atrium life-safety codes.

Project Team:

Owner:	The Taubman Company
Retail Tenant:	The Limited, Inc.
Architect:	Beyer Blinder Belle
Construction Managers:	Tishman Construction Co. (Owner)
	Gotham Construction (Tenant)

chapter twelve

Turning archaeological problems into assets

Sherene Baugher

Contents

Procedures and the law 136
Archaeological concerns 137
Choosing an archaeological consultant 137
The three phases of archaeological work 138
Phase 1: archaeologists in the library and on the land 139
Phase 2: site evaluation 141
Phase 3: the odds of your site requiring a complete excavation 141
The government(s) and you 144
The sunken ship 144
17 State Street 151
Conclusions 155
References 155

"We found a sunken ship!" Happy words for any archaeologist digging in a landfill site, but for most developers, one of their worst fears. A sunken ship in a landfill site means the archaeologists will need more time. Construction work could come to a halt; there might be lots of cost overruns; and the building's opening might be delayed. But that wasn't the case in New York City when, in 1982, an early eighteenth century merchant ship was discovered in a colonial landfill near Lower Manhattan's South Street Seaport.

There were no lawsuits! The construction was not delayed, even though the ship was fully excavated. The building actually opened two months

0-8493-7486-3/01/$0.00+$.50

ahead of schedule! There were even tax write-offs available to the developer for his archaeological expenses. Then the developer, his architect, and his project won city and state preservation awards.

A scene from a movie? No. Can it happened to you? Yes. You can turn an "archaeological problem" into an "archaeological asset." You can do this with good planning; an awareness of the local environmental regulations; and the help of the preservation and planning agencies within your community, whether you are in Boston, Chicago, London, Amsterdam, Ottawa, or even a small town.

This chapter briefly reviews the major steps you should take whether developers or your agents deal with the archaeology problem. I am focusing on two major case studies in Lower Manhattan, known in the field as the "175 Water Street Project" and the "17 State Street Project." The first illustrates how a developer can maneuver successfully through the regulatory process regarding archaeology. The second describes a disaster — the chaos and eventual legal penalties when there is a clash between a developer and the laws protecting archaeological sites. The two projects, while both from New York City, also take you through what are the typical procedures found in each of the fifty states, highlighting typical problems. And when appropriate, I cite other examples from beyond the boundaries of New York City to illustrate solutions and/or problems.

Procedures and the law

The need for archaeological work can be triggered by municipal, county, state, or federal laws. The developer usually has one of the firm's consultants (the architect, the landscape architect, or the construction manager) handle the environmental compliance issues. When a developer (or his agent) decides to approach the lead government agency to obtain building permits, he should immediately determine who handles the archaeological issues. Too often, within the maze of government bureaucracy, the developer is not told of the need to address archaeological concerns until well along in the review process and, sadly, close to the construction date. This 11th hour crisis can be avoided if the developer knows at the beginning of the review process what he has to do. Unfortunately, many times, the lead agency employees may not even be aware of all the steps within the compliance process. To save heartache and money, ask, "Who handles the archaeological issues? Is the archaeologist within a municipal or state agency?" Find out the name and phone number of this government archaeologist and make a personal contact. The government archaeologist should be able to explain the whole permitting process. Often the time frame of the permitting process provides ample time to undertake archaeological work well before any construction is scheduled to begin. For example, in New York City, the permitting process can take up to one year. If archaeological issues are addressed at the beginning of the process, the archaeological excavations can be completed long before construction begins.

Archaeological concerns

How do you find out if there are archaeological concerns regarding your property? Your first point of contact should be the State Historic Preservation Office (SHPO), they generally have files of previous archaeological projects. After meeting the archaeologist at the SHPO, there are other organizations you can contact. Historical societies may have data on the history of your neighborhood (but probably not on your parcel) that can give you some understanding of the neighborhood's importance and significance. In some suburban or rural communities, local colleges and universities have undertaken research on important privately owned local archaeological sites. For example, some sites are in farmers' fields. Inquire if any research has been done on your property by approaching the institution's archaeology program, but be sure to check with the history department, college archives, and any departments associated with architecture or landscape architecture. In addition, local government agencies, local historical societies, and public libraries may have copies of documentary studies and excavation reports of legally-mandated archaeological work already completed in your locale. It is possible that a neighboring property already has been the subject of an archaeological study, and it may contain general information on your neighborhood and perhaps even about your property.

Some government archaeologists have created maps that serve as predictive models for their city or county that highlight properties that have high archaeological potential. They can look at these maps and let you know if there might be archaeological concerns. In New York City, I often had developers come in to visit with me to inquire if their property was in an area that we had highlighted in our predictive model maps as having archaeological potential. Sometimes the developers would want this information even before they had purchased the property. At other times, they were debating the costs of going for a special permit (to build a larger or taller building than currently allowed) that would trigger all the environmental permits including archaeological evaluations compared to the costs of building "as of right" and thus not requiring any special environmental permits. The government archaeologist will provide you with the exact guidelines and forms you or your archaeological consultant will need to submit, and the subsequent steps you will be required to follow if your initial study indicates excavations are necessary.

Choosing an archaeological consultant

If there is an archaeological concern, how should you choose a consultant? Governmental agencies may provide a list of consultants that includes both individual archaeologists and larger firms. But you would do well to remember that not all archaeological consultants are equal in the quality of the competency of their staff or their reputation for careful work. Some archaeological companies will bid low intentionally so they can undertake the initial step in any archaeological study, the documentary work, knowing they will only prepare the sketchiest of reports. The review agency will reject the initial draft and require additional research so the report is thorough, and the low-bidding archaeological consultant will bill the construction company for cost overruns. Those overruns may end up costing you as much as your highest bidder! And in terms of a consultant's quality, the lowest bidder is not always the worst consultant and the highest bidder is not always the best consultant. Sometimes successful archaeological consultants bid lower because they have constant work, while some shoddy archaeological consultants try to make a killing at your expense. You might inquire with other construction firms about their experiences with archaeological consultants. You could inquire with the reviewing agency about the frequency of government-required rewrites by all your bidders, but especially by your lowest bidder.

Finally, it is important to realize archaeologists have specialties. Some specialize in underwater archaeology, some in American Indian sites, some in colonial sites, and some in military sites. Smaller consulting firms might subcontract to an archaeologist to cover something not within the field of any of their permanent staff. But occasionally these smaller firms, especially those with just one archaeologist, have been known to try to do all the work even when they lack the expertise. Thus it is important to find out about the specific expertise of the consultants, and whom they intend to hire as consultants, if any. If your property has a military site on it, you would do well to hire a consultant with expertise in military site archaeology. To excavate a Civil War site, for example, you wouldn't hire an archaeologist who was an expert on American Indian sites that existed before European contact because this archaeologist would have no expertise in the bullets, buttons, and artillery fragments sure to be found at any Civil War site, and because this archaeologist would lack an in-depth understanding of the nineteenth century documentary resources that would have provided adequate background information prior to any excavation. Unfortunately, when developers hire consultants without the proper expertise, their property can become embroiled in a major controversy over the inadequate handling of the site by the archaeological consultant. By hiring consultants with the appropriate expertise, controversies and construction delays can be avoided.

The three phases of archaeological work

The steps or phases involved in archaeological work actually work to your advantage both in terms of time and cost. The steps move from research in the documentary records and some preliminary probes in the field (Phase 1, relatively inexpensive) to carefully planned probes called field testing (Phase 2, more expensive) and then, in only a minority of cases, to a full-scale archaeological excavation of the entire site (Phase 3, also known as archaeological data recovery). In the vast majority of cases, you will not be required to undertake a complete excavation of a site (Phase 3) because one

Figure 12.1 Shovel testing in Ithaca, New York.

of the two earlier phases has indicated that no such large excavation will be necessary.

Phase 1: archaeologists in the library and on the land

Phase 1 is the first step in archaeological work. Phase 1 involves two separate undertakings, one in the library (known as Phase 1A or background research or documentary research) and one on the actual land (known as Phase 1B or archaeological inventory).

Phase 1A is undertaken in libraries and in other sources of documentation such as government archives. Phase 1A involves evaluating the available documentary literature to determine if the property might have an archaeological site. Phase 1A, the documentary research of Phase 1, is vital because a site's archaeological potential is based on two primary factors, both of which must exist at a site for it to be valuable. These two factors are the site's historical significance and the site's integrity. That is, even if a site was historically important, how much of the site is still intact and not disturbed by modern construction? Obviously, even if a site has historic importance (a sign at the place might read "George Washington camped here") the site will not have archaeological significance if George's campsite were replaced by the tunnels of an underground aqueduct or subway system! It is also important to remember that undertaking a documentary study (Phase 1A) does not automatically mean there is still an intact, relatively undisturbed archaeological site on your property.

Because Phase 1 (library and land) could be the only phase you are required to do, be sure that the archaeological consultant you hire is going to be thorough especially in Phase 1A, the library work. Library research is less expensive than excavations. Thus the archaeologist should be required to examine historic maps, aerial photographs, local histories, deeds, diaries, newspapers, collections of historic correspondence, insurance maps, tax maps, and building department permits to determine what was on the site and how the property changed over time. Sometimes historic paintings or photographs can help pinpoint the exact position of historic structures, and even the precise site of military camps. The consultant should also review previous archaeological reports for your neighborhood. Work nearby sometimes can help predict what you may find on your property.

Phase 1B (archaeological inventory) involves preliminary field testing carried out at the site (Figure 12.1). In Phase 1B, the archaeologist walks over the land looking for evidence on the surface — fragments of pottery, bricks, and other clues called surface finds. In Phase 1B, the archaeologist may also carry out relatively quick probes into the soil known as shovel tests and occasionally excavates a few small test squares, three feet by three feet or five feet by five feet (Figure 12.2). The archaeologist then records any evidence on a map.

Figure 12.2 Volunteers excavating on a New York City Landmarks Preservation Commission dig.

Phase 2: site evaluation

Based on the initial Phase 1 report by your consulting archaeologist, the government agency archaeologist charged with the legal supervision of your site will determine if you even need to proceed to Phase 2, known as site evaluation (Figure 12.3).

In Phase 2 your archaeological consultant will evaluate the significance of the site(s) found in Phase 1. Phase 2 involves both additional library research and fieldwork in order to determine if more intense fieldwork (Phase 3) is required. This research (Phase 2) may be undertaken because the archaeological evidence in Phase 1 raised issues or questions no one anticipated. For example, a nineteenth century farmstead which the archaeologist knew was at a site because of an historic map may also turn out to have an American Indian town site buried beneath the farm. Finally, if the reviewing agency determines the site to be significant (based on the Phase 2 work), then Phase 3 will be required. Phase 3 can involve either a full-scale excavation or the preservation of the site.

Phase 3: the odds of your site requiring a complete excavation

New York City, by requiring extensive documentary studies, exemplifies what we might call the odds. During the period 1980-1990, only 50% of those documentary studies found the historic sites still intact, meaning that only half the sites required any archaeological field testing whatsoever, including shovel tests. Because archaeological consultants conducted extensive work in the archives of the city and in the libraries of historical societies and museums, documents often demonstrated that the site had already been destroyed by late nineteenth century or twentieth century construction. Having done thorough documentary studies in every phase of the work, developers avoided Phase 3, the most costly, full-scale archaeological excavations, 90% of the time. But the projects that represented only 10% of the total, and were fully-excavated as "Phase 3" sites, were truly outstanding: a Dutch trading post, colonial Dutch and English homes, shops, taverns, and the early eighteenth century sunken ship mentioned earlier! The economics lesson is as clear as the archaeological one: at each stage of the process, thorough documentary work saves time and money.

It is amazing to realize what can survive in the archaeological record. A parking lot may serve as a sealant over an archaeologically rich site: a time capsule under asphalt. In Lower Manhattan, the foundations of seventeenth century Dutch buildings lay underneath parking lots and even under nineteenth century foundations of buildings that were four and more stories high (Figure 12.4). Therefore it is all the more important to make sure the consultant you hire undertakes a thorough documentary study to determine: not only if the property could contain a significant site; but more important, to determine if the site still exists. Some states require a very minimalist background study and often government archaeologists will, based on scanty evidence, require field testing. To undertake a thorough historic study will cost more up front, but in the long run, it will be far less expensive than undertaking unnecessary archaeological fieldwork once your project is underway.

Figure 12.3 City Hall Park, New York. (Courtesy Carl Forster.)

When your archaeological consultant recommends (to the review agency archaeologist) that field testing is needed, it is important to know why that recommendation is made. If very thorough research has been completed, the archaeologist should be able to focus on those sections of the parcel that still may contain an intact archaeological site. The archaeological consultant should be able to plan an efficient field testing strategy that is time and cost-efficient, and meets professional standards. The documentary work should enable a consultant to target those specific parts of the property still containing intact, undisturbed archaeological material. Above all, the consultant should not have a "one size fits all" testing plan (and sadly some consultants do), and a consultant's specialization in one aspect of archaeology (American Indian, colonial, etc.) should be considered before that consultant is hired. The way an archaeologist would discover whether the approach to locate the exact boundaries of a twelfth century Native American town site is different from the approach an archaeologist would take to find a nineteenth century Euro-American fortification. The archaeologist would have to be familiar with the desirable environmental setting for the Indian town, such as a flat, elevated, well-drained area near water. The site of an Indian town might initially be indicated by surface finds of pottery fragments on top of a large hill. Because no twelfth century Native American map exists, the archaeologist uses a series of shovel tests, excavation squares, and trenches to narrow in on the town's location. With regard to the Euro-American fortification, however, careful research in local libraries should uncover maps and/or other documents locating the approximate or even exact location of the fortification. The differences in specialties, however, are only obvious once you know differences exist.

Figure 12.4 17th century foundations. (Courtesy Carl Forster.)

Developers who see only "generic" archaeologists may find the results costly. For example, in the 1990s, in Gettysburg, Pennsylvania, a national food chain intended to build a huge store on what turned out to be part of the site of the Civil War military hospital, Camp Letterman. A controversy, that perhaps could have been avoided, soon erupted.[1] The developer hired an archaeological consultant who had no experience in military sites. The site reflected an especially poignant chapter in the history of the battle of Gettysburg because the hospital at Camp Letterman served thousands of wounded men, both Union and Confederate. Civil War photographs showed precisely where the temporary hospital tents and other structures of Camp Letterman were, and some of those photographs had recently been published in 1995 and 1997.[2,3] But unfortunately, a minimalist historical study was undertaken, and these valuable resources were not used in planning the fieldwork. Subsequently, preservationists and archaeologists raised questions over whether the fieldwork at the Camp Letterman site was adequate and appropriate.[4] The controversy delayed the project. All this could have been avoided if the developer had required the archaeological consultant to undertake a thorough documentary study before any on-site excavations were begun.

The government(s) and you

A government archaeologist (local, state, or federal) will evaluate the archaeological reports drawn-up by the consultant you have hired. These reports are known as cultural resource management or CRM reports. Your consultant's report should present clear and justifiable recommendations for or against additional archaeological testing. If your consultant believes additional fieldwork is necessary, your consultant should also submit to you and to the government archaeologist detailed scopes of work that clearly define the exact nature and location of the testing, and the projected number of people and days of work involved. However, it is the government archaeologist who will make the determinations if additional phases of work are necessary, and will also determine if your consultant's testing strategy is appropriate. The government archaeologist will ultimately approve or reject all phases of the work. Here are two case studies to take you through a project and illustrate the various problems, complications, and successes.

The sunken ship

In 1981, Howard Ronson, a British developer, requested a discretionary permit from the New York City Planning Commission so he could erect a skyscraper at 175 Water Street, on the eastern edge of Lower Manhattan. The colonists had named the street Water Street because it ran along the shore of the East River, whose swift current flows between Manhattan and Long Island. But by the late twentieth century, colonial and nineteenth century landfills had left "Water Street" several blocks inland. A copy of the permit granted to Mr. Ronson was sent to the City Archaeology Program, a division of the New York City Landmarks Preservation Commission. At that time, in my role as the city archaeologist, I evaluated its archaeological potential. Fortunately (and at the expense of the taxpayers, not developers), my staff and I had already helped complete an historical and archaeological study for Lower Manhattan, so there was no question the block slated for development by Mr. Ronson was significant in terms of the city's early history.

Maps and documents indicated the site contained merchant shops and residences from the late eighteenth and early nineteenth centuries. The question was whether those sites were still intact. The developer hired a consultant, Soils Systems, to undertake the documentary research. In 1981, the parcel was a parking lot, but previously it had contained mid-to-late nineteenth century buildings with very shallow foundations. Based on the depth of buried seventeenth and eighteenth century archaeological remains found on other previously excavated blocks in Manhattan, it was assumed that this block would also contain buried colonial structures and associated artifacts. The question was when and how to undertake the archaeological work. Lawyers for both the developer and the city debated and debated the issue. As we listened to lawyers in meeting after meeting, the project architect and I began to feel as if this project would never get off the ground. Then, after one meeting, the architect, Robert Fox of Fox and Fowle, made a novel proposal: He suggested that we should send the lawyers away, and the developer's architect (Mr. Fox) and I could resolve all the conditions for the archaeological fieldwork so the site could be excavated professionally and the office tower could be built in a timely manner. Both the developer and the head of the Landmarks Commission thought this highly unusual approach might work, but they also both agreed that if this "gentlemen's agreement" didn't work they would call the lawyers back. Under my supervision, as the city archaeologist, a cultural resource management (CRM) firm and their archaeologists worked from mid-October to mid-January, and uncovered almost 250,000 artifacts from the eighteenth and early nineteenth century.[5] Near the end of the archaeological dig, the developer signed an agreement that if a snow storm delayed the excavation by one or more days, an equal number of days would be added to the dig's schedule to make up for the missed work. Because we knew the site originally had been part of the East River, the developer also agreed that if we found a sunken ship, the conditions and length of the dig would be renegotiated. A few days after signing the agreement we had a major snowstorm. No problem, the developer agreed to add some extra days to the excavation. The dig ran smoothly, months in advance of the March 4th starting date for construction.

During the last days of the dig in early January 1982, the archaeologists found what they thought looked like a section of a sunken ship. We knew the developer's parcel was on man-made land. In the seventeenth century, that land was still part of the East River, and in the early eighteenth century wharves extended there from the shore. By the 1750s, colonial developers bought the water "block" and began filling it in with dirt (and all sorts of colonial garbage). However, the extensive and thorough documentary research had failed to uncover any record of a ship being sunk as landfill on this particular parcel. Later, additional documentary research was undertaken by Warren Reiss for his doctoral dissertation on our sunken ship. However, just as the previous researchers, Reiss also was unable to uncover any specific documentary evidence regarding the sinking of any ship at the 175 Water Street project site.[6] So we assumed that the archaeological field team had located a wharf. We enlarged the trench and brought in a maritime historian from the South Street Seaport Museum to inspect the wooden structure. He was delighted to tell us that indeed we had found a sunken ship. Not just any ship, but a large, bulky eighteenth century merchant ship. And as we widened the excavation, there it lay in the mud, from bow to stern! This became the first time an early eighteenth century merchantman had been uncovered by archaeologists. The previously excavated eighteenth century ships were men of war. Thus, our sunken ship provided a rare opportunity for historians and archaeologists to study this type of vessel (Figure 12.5).

Luckily, we had a written agreement that allowed us to renegotiate the time frame if we found a sunken ship. To our delight, the developer was interested in maritime history and he willingly agreed to fund the ship's excavation. He brought in a large team of maritime archaeologists from all

Figure 12.5 One half of the ship was under Front Street, where the taxis are parked. (Courtesy of Carl Forster.)

Figure 12.6 Grid frame for the ship. (Courtesy of Carl Forster.)

over the country to undertake the excavation. The ship was partially under a city street (where it remains to this day) and partially on the developer's property. The goal was to excavate fully the section of the ship on the developer's property. Because the developer's property cut across the length of the ship at an unusual angle, more than half the ship's hull, including the entire bow and a part of the stern, was excavated. Enough of the ship was visible to obtain all the necessary information about the vessel's construction and use. The archaeologists removed (timber by timber) the first twenty feet of the ship including the bow, in the hopes that the bow could be conserved and donated to a museum (Figure 12.6). (Part of the ship, still packed with colonial mud and debris, remains at vigil, supporting a part of the modern street).

By February 1982, both sides agreed that some construction preparation could take place on the half of the block that had already been excavated to everyone's satisfaction. Both sides also felt that the site preparation work for construction would not interfere with the excavation of the ship and would pose no safety hazards to the archaeologists. At this point, the site construction manager interacted with the archaeologists on a daily basis to insure there were no safety problems or any problems that would interfere with the archaeological excavation. This was an example of both sides working together for the mutual benefit of both.

The archaeologists agreed to work longer hours (ten-hour days) and six day work weeks to finish the dig as quickly as possible yet still maintain high professional standards (Figure 12.7). Both sides were still aiming for the original March 4th start-up date for construction (the date established prior to the discovery of the sunken ship). The developer for his part provided lab space and storage space in his construction headquarters across the street from the site. He even had hot lunches brought in for the archaeologists. Again, these are examples of both sides willing to do more than was required by the law.

The archaeologists regularly gave tours of the site to Mr. Ronson's clients, but at first the site was not open to the public. However, with the enthusiastic support of Mr. Ronson, Kent Barwick, the New York City Landmarks Preservation Commissioner, arranged to open the site to the public. On the last Sunday in February 1982, the site was open from 10:00 a.m. to 6:00 p.m. Temporary exhibit panels were set up on the property fence, and tours were given all day. Amazingly, over 10,000 New Yorkers toured the site, and by mid-day people were waiting on a line four blocks long to see our ship. The public interest and enthusiasm for the ship amazed everyone. The ship provided positive publicity for the developer and his project. The city's newspapers, television news, and other media gave the ship great publicity, and the ship even achieved primetime on the national television news (Figure 12.8).

The spirits of everyone were extremely high as we neared March 4th, the date when construction was due to begin. On March 3rd, the archaeological team agreed to work through the night the finish the dig on time. At 6:30 a.m.,

Figure 12.7 The bow of the ship at 175 Water Street. (Courtesy of Carl Forster.)

March 4th, with a half hour to spare before the first construction worker was due to arrive, the last timber from the ship's bow was removed and the excavation was completed. Everyone was thrilled and proud at both the dig's high professional quality and its on-time completion. We felt we demonstrated that professional rigor could be maintained while still working successfully within the time constraints of a construction project. Mr. Ronson hosted a party for the archaeologists at the elite Waldorf Astoria Hotel to celebrate the on-schedule and successful completion of the historic excavation.

The wooden bow of the ship was placed in a warehouse, awaiting analysis and conservation. We hoped that a maritime museum would undertake the conservation. When it became apparent the cost of conserving the bow was preventing any museum from accepting it, developer Ronson graciously agreed to provide the $350,000 needed to undertake the conservation. Again, this is another example of the developer going far beyond any legal requirements to complete the project successfully. After the conservation was completed, Mayor Edward Koch held a press conference to encourage a museum to come forward and accept the ship's bow.[7] Eventually, the ship was donated to the Mariners Museum in Newport News, Virginia, at the southern end of Chesapeake Bay. The ship's return to the Chesapeake was appropriate, because a laboratory analysis of the ship's timbers had enabled archaeologists to determine the ship had been built somewhere in the Chesapeake Bay area. Thus when our ship took its place in a museum on Chesapeake Bay, the ship was returning to where it had been launched — its birthplace. Other fascinating details of ship's history were also revealed by

Figure 12.8 Labeling at the ship's timbers at 175 Water St. (Courtesy of Carl Forster.)

laboratory analysis. Wormholes below the waterline of the ship's hull indicated that it also sailed the Caribbean before it was purposefully sunk as landfill in colonial New York.

The extraordinary cooperation on the 175 Water Street project was recognized by a series of awards. The Municipal Arts Society awarded its preservation award in 1982 to the project and to Mr. Ronson. Then Governor Hugh Cary of New York presented the State Preservation award to me as the City Archaeologist, and to Robert Fox, Architect, for our leadership on this project. The 175 Water Street project was truly a turning point in national urban archaeology, demonstrating to the nation that archaeology meeting high professional standards could be carried out on a legally mandated project without incurring any construction delays. In fact, the building was completed two months ahead of schedule. As an added bonus, if an archaeological collection were donated to a museum, historical society, or university, the IRS (in the 1980s) allowed developers to use the cost of the excavation as the dollar value of the donated archaeological collection. With changes in tax laws, accountants should investigate if there are still any categories in which archaeological expenses can be deducted from business taxes.

Figure 12.9 17 State Street and Seaman's Church. (Courtesy of Carl Forster.)

17 State Street

Today, the 17 State Street building stands at the southwestern tip of Manhattan, facing the Hudson River near South Ferry. This towering building stands as a visible reminder of what can go wrong. The agents for the 17 State Street project developer requested a discretionary permit from New York City and proceeded to go through the same layers of bureaucracy that had led to a successful and harmonious process at Howard Ronson's 175 Water Street. But instead of uncovering a sunken ship and serving as a national example of positive preservation, this project served as an expensive example of what not to do (Figure 12.9).

The Department of City Planning required an environmental review at 17 State Street. Just as at 175 Water Street, the 17 State Street site was in Lower Manhattan, and was also in an area already flagged by the City Archaeology Program as having archaeological potential. The New York City Landmarks Preservation Commission (home of the City Archaeology Program) advised the planning commission that the site of the proposed new office tower potentially contained archaeological material from the seventeenth and eighteenth centuries. The developer's agents were informed that an archaeological documentary report was required to determine if the site still contained any undisturbed material.[8] The developer's agents obtained an "as-of-right" building permit while the environmental review was pending. They excavated the site for the new building's foundation, and destroyed any traces of the potential archaeological site.[9] This loophole has been plugged now. Now, before issuing a construction permit, the building department checks to see if the project is also undergoing any review for a discretionary permit. However, a documentary study required by the planning commission and paid for by the developer showed a portion of the project area had been relatively undisturbed and was identified as having archaeological potential prior to the developer's destruction of the site.[10] Abraham Isaacs, a merchant and member of New York's Jewish community lived on this property from 1728 to 1754, and this site could have been the first eighteenth century Jewish home excavated in the northeast. Thus the property had archaeological significance for both New York and the whole northeastern part of the United States.[11]

The 17 State Street site's destruction clearly challenged the enforcement of the city's environmental review regulations. The environmental review process would be undermined seriously if a developer could destroy an archaeological site while the project was undergoing an environmental review. This became a test case and went before the City's Board of Standards and Appeals (the body that, at that time, resolved conflicts between applicants and the planning commission). Because the developer destroyed a significant archaeological site, the Landmarks Preservation Commission requested some form of mitigation. Representatives of community organizations and professional groups, including the Professional Archaeologists of New York City, appeared at public hearings to oppose the developer's position.[12] The developer's battery of attorneys fought hard to avoid any penalty, but failed. The Board of Standards and Appeals decided in favor of mitigation and a mini-museum had to be included in the developer's plaza.[8]

Under a legally-mandated memorandum of agreement, the developer had to pay for the design and installation of a mini-museum in the plaza of his new building (Figure 12.10). He also had to fund its maintenance, management, and public education programs for five years. In 1989, the developer was allowed to turn over the administration and operation of the exhibits to the South Street Seaport Museum. The New York City Landmarks Preservation Commission had to approve the completed space of the mini-museum and the installation of the exhibits before the developer was able to obtain a permanent certificate of occupancy for the building.[8,13]

Figure 12.10 Museum at 17 State Street site. (Courtesy of Carl Forster.)

The mini-museum at 17 State Street, entitled New York Unearthed: City Archaeology, opened in the fall of 1990 (Figure 12.11). This museum is a small building in the public plaza of the 17 State Street building, and is connected to the office tower by a tunnel. The museum has two floors of exhibit space: about 400 square feet at the plaza level, and some 1,200 square feet at the lower level.[14] The costs of design, construction, and installation of the little museum, plus its five year operating budget, far exceeded the cost of any New York City archaeological excavation undertaken between 1980 and 1990, but the mini-museum can never replace what was lost.

Figure 12.11 Another view of the museum at 17 State Street. (Courtesy of Carl Forster.)

Conclusions

Archaeological work need not be a stumbling block. If you know the process, things can work smoothly. Actually, archaeological discoveries can generate positive publicity for a project. Even a complex project such as the ship did not delay construction. During the building boom of the 1980s, the rigorous archaeological standards for documentary work and field work never delayed construction projects, not even for a day. People in the construction industry need not fear archaeology.

References

1. Editorial, Hallowed ground in jeopardy: the latest battle of Gettysburg., *The New York Times*, A18, July 4, 1997.
2. Coco, G.A., *A Strange and Blighted Land – Gettysburg: The Aftermath of Battle*, Thomas Publications, Gettysburg, 1995.
3. Patterson, G.A., *Debris of Battle: The Wounded of Gettysburg*, Stackpole Books, Mechanicsburg, 1997.
4. Files on proposed development for Camp Letterman, *Hospital Woods*, Files at Gettysburg Battlefield Preservation Association, Gettysburg, 1997.
5. Geismar, J., *The 175 Water Street Project*, Report on file with the New York City Landmarks Preservation Commission, New York, 1983.
6. Reiss, W., personal communication, 1988.
7. Greer, W., City seeks help for a homeless ship, *The New York Times*, A20, December 23, 1984.
8. New York City Landmarks Preservation Commission, *17 State Street project*, Files at the New York City Landmarks Preservation Commission, New York, 1986.
9. Clifford, T., Piece of city's history buried at building site, *Newsday*, 9, July 27, 1986.
10. Geismar, J., *17 State Street: an archaeological evaluation, Phase 1 documentation*. Report on file with the New York City Landmarks Preservation Commission, New York, 1986.
11. Vollmer Associates, *Final environmental impact statement for the proposed project at 17 State Street, New York City*, Report on file at the New York City Planning Commission, New York, 1986.
12. Wall, D., Comment for the public hearing on the draft impact statement for the proposed construction of the 41-story office building located on Block 9, Lots 7, 9,11, and 23; 17 State Street, Manhattan, CEQR no. 85215M, BSA no. 53285BZ, Board of Standards and Appeals Chambers, *PANYC Newsletter* 30, pp. 8-9, July 9, 1986.
13. Woodoff, J., Proposed mitigation plan for 17 State Street, Cover letter and plan submitted to the Honorable Sylvia Deutsch, Chair, New York City Board of Standards and Appeals, 17 State Street files at the New York City Landmarks Preservation Commission, New York, August 22, 1986.
14. Baugher, S. and D. Wall, Ancient and modern united: archaeological exhibits in urban plazas, in: *Presenting Archaeology to the Public: Digging for Truths*, Jameson, J. Jr., Ed., Alta Mira Press, Walnut Creek, CA, 1996, pp. 114–129.

chapter thirteen

Trees in urban construction

Jason Grabosky

Contents

Basics 158
Engineering a viable tree rooting zone 159
Media specification 160
Berms or raised bed planters 162
Vegetation strips (tree lawns) 164
Containerized systems 167
Media design considerations 167
Rooftop or setback planting 169
Containerized plants and planting beds 170
Estimation of water use and soil volume 170
The tree pit as a dysfunctional design 171
Alternative street tree-pavement systems 172
Cantilevered pavements 172
Combination container-cantilever systems 173
Amsterdam tree soil 174
CU Soil® 174
Designing solutions 175
  Berms or raised planters 175
  Vegetation strips and inter-connected tree pits 177
  Rooftops and setbacks 178
  Containerized planting 180
Protection of trees during construction 180
Designating the protection zone 183
References 188

0-8493-7486-3/01/$0.00+$.50

Basics

Trees have recognized values and benefits in urban construction. Many economic, environmental, and aesthetic benefits are linked to healthy trees in urban areas. As the largest living component of the urban landscape, trees increase in value as they grow. The challenge lies in establishing trees successfully in the restricted spaces typical of urban environments. Trees can create problems when roots or branches interact with structures and utility corridors. Such problems can render the best-intended landscape design a detriment to long-term success of a project. This chapter will address current research and strategies being implemented successfully to establish trees in restricted urban spaces while minimizing the structural problems of plant success.

The impact of trees on the environment increases with their canopy.[1] The collective forest canopy can reduce temperatures in the urban "heat island"[2,3] while they filter airborne pollutants and particulate matter that would otherwise lodge in our lungs.[1,5] Well-placed trees and vines reduce energy costs in heating and cooling on an individual scale.[5-7] Property values and parking behavior are positively influenced by healthy trees.[8] The seasonally dynamic appearance of trees impart an aesthetic value along with a reduction of glare from smooth surfaces and the perception of noise reduction.[9-11]

Trees evolved in forest groups, not as individual specimens in modern cities, and that produces a series of challenges in establishing urban landscapes. The transition from forest to city alters tree structure and biological functions. Trees evolved with competition for light access, water, and mineral resources. Few trees naturally survive to be the dominant individuals within their respective canopy level. In urban situations, where each tree is an investment, survival is essential.

Species and cultivar selection can be used to adapt a tree planting for above-ground space restrictions (canopy structural stability, environmental hardiness, light patterns, building over-hangs, or utility wires). Construction design should give attention to the below-ground requirements and restrictions for long-term establishment of the tree. Tree roots need soil to provide oxygen and water in balance, within root-penetrable voids. There is general agreement that water is often the currency for transplant establishment and success.[12-15]

Trees require large amounts of viable soil to meet their water and nutrient demands. Water use can be estimated from water use data from other plant species and climatological data. One calculation estimated water use of 1.1 $gal \cdot ft^{-2}$ per week (0.94 $l \cdot m^{-2}$ per week) for a mixed shrub-tree planting of medium planting density at moderate water demand (see R. Harris, 1999 pp. 378-390 for formulae,[16] and the Weather Bureau for climate data). J. Roger Harris estimated 10 gallons of water per 3 day irrigation cycle for newly transplanted trees (in Virginia) with a canopy diameter of 8 feet.[15]

Irrigation and fertilization systems are used to compensate for reduced soil volumes and the stresses of the developed site. Aerobic microbial processes and associations are also present within the soil. Soil is fragile from a horticultural sense, and maintenance personnel often cannot repair damage done to a soil, especially after the plants have been installed. In restricted urban planting spaces, proper media selection by a qualified horticultural consultant during project design with root zone protection during and after installation is essential.

Trees in "downtown" urban environments often never grow to their projected size in the design plan.[17-20] Life expectancies have ranged from less than 7 to 20 years from time of transplant, depending on the defining location parameters. Exceptional urban forestry programs and benevolent climates can experience upwards of 60 years in the same situations. Below-ground limitations are a primary cause for the general "failure to thrive." While trees can live from 80 to well over 100 years, a life expectancy equal to the cycle of urban renewal should be the minimum goal for design. Limited soil volume from soil contaminates, debris, pavement section materials, and compaction of soil are problematic in designing or establishing trees in urban situations. Where tree growth is vigorous, roots are associated with damage to surrounding structures; heaving pavement, or disrupting foundations and utilities.[20-26]

Careful plant selection and hardscape design can produce success for both trees and structures. Foundation subsidence is largely a function of tree water extraction on reactive clay soils, and guidelines for planting distance from foundations range from 10 to 20 meters depending on the tree species.[27] While new sewer designs can delay root ingress into the utility line,[24,25] the spacing criteria for foundations can also be used to delay root-sewer conflicts. Danish research found ivy on walls not to be as detrimental to wall materials (measured by moisture levels and integrity of the stone and grout materials) as once thought, although there is a problem with breakage of gutters and downspouts with radially expanding branches.[5] Load bearing media are being used for plants in high use natural or park areas, or in places where pavement and root zones overlap.

Engineering a viable tree rooting zone

Soil selection (or media design) is crucial for project success. Soils provide water reserves for the tree, nutrients for growth, air for root respiration, anchorage for roots to support the tree under a load, biological associations related to root functions, and other biological activities related to nutrient mineralization. Estimates for the amount of soil needed for urban tree establishment have ranged from a minimum of 3.7 cubic yards[28] to over 30 cubic yards with a suggested maximum depth of 2 feet.[16] (Figure 13.1.)

Media are specified relative to the system wherein the tree is planted; e.g., open areas for tree establishment such as parks, berms (Figure 13.2), or median strips (Figure 13.3) will use a specified agricultural-like soil. There is no legal definition of topsoil, so careful definition is important for tree establishment success. Containerized media design addresses aeration, irrigation, and nutrient issues particular to the lateral and depth limitations of

Comparative Geometry of root volumes Note: Many street tree planting projects are based on 40 foot centers.	Volume observations and recommendations from several published studies
4cubic yards 6 x 6 x 3feet 8 cubic yards 6 x 12 x 3 feet Bakker 30ft dia. crown 65.4 cubic yards 12 x 49.1 x 3 feet Lindsey 30ft dia. crown 52.3 cubic yards 12 x 39.3 x 3 feet	Urban [33] 3.7 yd^3 minimum for survival Arnold [35] 8.3 yd^3 for 21-40 foot tree Kopinga [36] 92.6 yd^3 for large tree Perry [37] 1.0 yd^3 for every 1 inch caliper 22.2 yd^3 for 10 inch caliper Moll et. al. [38] 44.4 yd^3 for 25 inch cal. tree Bakker [39] 2.5 ft^3 per ft^2 crown projection Lindsey [40] 2.0 ft^3 per ft^2 crown projection

Figure 13.1 Estimations of soil volume requirements for tree establishment and survival. Estimates include observational data, minimums from field experience, and methods of predictive estimation.

Figure 13.2 Berms can be used for screening or to gain soil volume in narrow spaces designated for planting. (Photo credit: N. Bassuk.)

a confined, or closed system. Root zones that need to project below paved surfaces have the additional requirement of bearing capacity for durable pavement design. Such load-bearing root zone media are at the cutting edge of urban tree media design. Correctly executed, there are possibilities to integrate durable pavement design and tree root zones for pedestrian malls, parking lots, or sidewalks where there is no option for an expanded, non-paved root zone.

Media specification

Plant survival requires soils with a balance of air-filled and water-filled soil pores. It is more efficient to kill trees by drowning than by droughting. Roots

Figure 13.3 Median strips can be used for aesthetics, tree establishment in divided traffic corridors, or parking lots. (Photo credit: J. Grabosky.)

must have oxygen, so positive drainage must be provided to the root zone. Water must be able to infiltrate, move within, and drain away from the desired active rooting zone. Drainage design must address infiltration, surface runoff, internal drainage and movement within the root zone, and deep drainage out of the planted zone. On urban sites, drainage is influenced by soil compaction, interfaces between dissimilar materials, and existence of rubble within, below, or surrounding the designed planting space.

Topsoil specifications should be designed to meet horticultural needs in relation to the existing conditions and design limits on a construction site. A qualified soil scientist, or consulting arborist with appropriate training should be used to properly define a soil specification acceptable to regional needs. The language of the soil specification should be precise with mechanisms of approval-rejection clearly stated. The body of the specification should meet regional standards while addressing several specific topics:

- Definition and delineation: Placement and protection language is necessary, but quality control is out of the supplier's control after delivery of the materials. The installation contractor needs to accept final responsibility for protecting the investment by careful handling and protection of the soil during the project.
- Physical characteristics: sections specifying particle size distribution in sieve gradations, D10, D60, D90, or soil classification; organic matter content; and soil structure, if testing facilities are available.
- Chemical: Regional soils will dictate acceptable ranges of suppliers and should match the requirements of the plant material. It is easier and more efficient to match the plant to the site. pH (plants typically prefer 6.0–7.0, but can vary by species, and some tolerate elevated pH (up to 8.3). Amending the soil is an option requiring maintenance of

pH level. The carbon to nitrogen ratio (particularly in the organic compost section, total carbon to nitrogen in system should be <20:1 and no more than 33:1). This is related to mineralization and availability of nitrogen to the plant. Cation exchange capacity (influences nutrient availability) often in the recommended range of 5 to 15 meq/100g. Electronic conductivity (<2 dS m^{-1}) or soluble salts (20 ppm) Sodium Adsorption Ratio (more of an issue in the west and southwest, but also in heavily salted pavement corridors where plants are established nearby 12 or more is problematic for plants, <8 for specification).

- Amendments: Composts (a separate specification of acceptable materials, a maximum of 5% by weight which will still result in some settling); sands (angular coarse sand only, at least 75% >0.25mm in size, should be used with composts); chemicals; fertilizers; anything to change the physical or chemical nature of the soil. Include definitions, acceptance testing, and rejection criteria.
- Environmental statements, if necessary: Such as prohibitions on peat moss for environmental protection, or restriction on soil source materials to preserve soil or prevent stripping of prime farmland.
- Delivery and acceptance conditions: Moisture control and density; screened soil material, or bulk.
- Deleterious materials: General statement prohibition of physical and chemical contaminates, pesticide history on source material, trash.
- Storage and holding: For the contractor and installation phase. Protection from irrigation or rain, traffic or over-working of material; protection of the material from site construction, compaction, or contamination once in place.
- Post-installation testing and acceptance (rarely done, but should be required): This section should include density, consistency, depth, drainage, and grade.

Many urban planting designs deal with restricted soil volumes in contact with pre-existing soil materials or fill. Generally, greater similarity between the soil material added and the existing soil underneath results in better water movement between the layers. This is important for planning drainage. A coarse granular medium placed under a compacted soil high in fines content will not provide effective natural drainage out of the fine material rooting zone. Soil specifications are subject to regional applications (usually a coarse sandy loam material with 5% organic material). If an organic compost is used, the carbon to nitrogen ratio should be less than 20:1. Organic mulches should be composted prior to installation.

Berms or raised bed planters

Berms gain soil volume above-grade for noise reduction, visual screening, and/or horticultural advantage (Figure 13.4). Irrigation and nutrient

Figure 13.4 Raised containers increase root zone volume in spatially limited situations. In this case, the containment sides are used for seating. The geraniums planted in the tree planter are watered and fertilized. The tree benefits from this increased maintenance, compensating for the limited root zone. (Photo credit: N. Bassuk.)

management protocol will be dictated by the plant material and soil materials used in creating the berm, and should be addressed in the design phase by the consulting engineer, soil scientist, and arborist. Annual flowers could be installed to encourage shop owners or community groups to irrigate, fertilize, and maintain the berm for their own benefit. The trees will also benefit by the irrigation and cultural attention. Drainage and planting viability are influenced by the soil material used, compaction levels used in construction, and the geometry of the berm.

Compaction of the soil should be discouraged for planting purposes, but placed fill materials will require some level of minimal compaction to prevent settlements and lodging of trees planted on a physically unstable soil mass. Some advocate compaction to 80% proctor density. Let irrigation do the settlement if possible, or only compact enough to eliminate anticipated visual settlements (clairvoyance counts in the latter case). Saturated hydraulic conductivity of the soil once in place should be 1 to 6 inches per hour at saturation; much higher will result in increased irrigation and less could be limiting in aeration status of the root zone. Installation of the fill should minimize the interface of the two layers. One method is to spread several inches of the fill over the existing soil and till the layers together, then proceed with additional lifts of the new fill material.

The size of the berm is determined by site constraints, purpose of the berm (noise control or strict aesthetics), and the size of the plant material to be established. The geometry of the root zone should accommodate the development of a root plate for tree stability during wind events when designing a berm's width (Figure 13.5). German tree protection standards, based on observational data, estimate a minimum root plate radius dimension of 2.7 yard (2.5 meters).[34] British ordinances for protection require a

Figure 13.5 The width of a berm or any other root zone must factor in the width of a minimal root plate for stability and avoidance of root-pavement problems: A fact lost on the author of this planting design. (Photo credit: E. Gilman.)

radius of 4.4 yards (4 meters) for a 10 inch (25.4 cm) caliper tree.[35] The final shape and size of a root plate zone will be influenced by prevailing wind directions and the designed size of the tree at project maturity. Root excavation data has suggested that tree roots will grow (in penetrable soil) first in response to water availability, and then parallel to wind direction for mechanical stability. The material's angle of repose should be considered in the design of the berm sides. If the berm is to be mowed, consider a 1:3 maximum gradient for mower safety. Unfortunately, urban spaces are limited in size, often precluding a free-standing berm with gently-sloping sides. Retaining walls can be used to stabilize or define the berm or raised bed while integrating pedestrian seating elements under the designed tree canopy (Figure 13.6).

Vegetation strips (tree lawns)

Vegetation strips parallel to the traffic corridor can be used to establish trees, provided the soil within the vegetation strip is not compacted from the construction of pavement, curbs, and sidewalks on either side. The material could be excavated and replaced by a specified "top soil," but must be to at least 18 inches to provide the soil necessary for the tree. Drainage of those panels is required to prevent trapped water from having an impact on the surrounding pavement. Utility corridors should take precedence over tree selection or establishment when they occupy the same space and creative design cannot integrate planting and utility needs. While many small tree species can be selected to prevent utility line interference, and newer sewer designs can minimize root ingress, root zones cannot always fit into the limited space within the public right-of-way. Narrow vegetation strips (less

Figure 13.6 Tree in raised planter with seat wall, University of Florida, Gainesville campus. A dedicated irrigation line is supplied to planter. (Photo credit: J. Grabosky.)

than 10 feet) eventually provide a functionally two-dimensional shallow root zone parallel to the street, unless the roots get under the sidewalk (Figure 13.7). The narrow root zone can lack structural stability for wind-loading perpendicular to the street, causing wind-throws. Narrow berms can suffer the same fate, however the soil volume and mass added by the depth of the berm and side containment structures can provide some additional stability. Setback ordinances[36] to place trees into residential properties can be used where free soil access is available opposite the sidewalk. This gives many options for tree size and establishment success, given the additional space above and below-ground.

Tree vigor in narrow vegetation strips can be improved if the root system can move under the pedestrian corridor, and into adjacent soil beyond the pavement section (Figure 13.8). Small trees of less than 18 feet (5.5 meters) tall would be obvious choices for small spaces, planted at least 3 feet from the edge of the pavement to allow buttress root growth without immediately lifting the pavement (Figure 13.9). Root barriers have been suggested as a method of preventing roots from lifting the pavement. The materials have varied in effectiveness in research trials demonstrating installation technique, depth of penetrable soil, compaction levels within the vegetation corridor, and provision for radial tree root-plate growth, warrant consideration. Materials in use range from rigid structural plastics to herbicide enhanced geo-textile panels. One untested use of barriers would be to train

Figure 13.7 Narrow tree planting zones pre-dispose the design to pavement damage if the trees are successful in surviving and growing. (Photo credit: E. Gilman.)

Figure 13.8 A tree's vigor can be increased in many cases by moving the tree to greater soil volumes on the other side of the sidewalk. (Photo credit: E. Gilman.)

tree root systems into conduits below the sidewalk using root barrier materials to guide roots to the conduit-encased escape channels under the pavement into free soil beyond the pavement enclosure.[37] Strip drains might accomplish the same root direction training objective while providing drainage for the sidewalk base material, but these systems need testing and field evaluation trials.

Figure 13.9 Trees and pavement grow older together, but as the tree grows, the concrete simply ages. Time, traffic, and the soil materials under the sidewalk influence the damages caused by root growth under the pavement. (Photo credit: N. Bassuk.)

Containerized systems

There are sites without contact to the existing soil, such as totally containerized systems and larger beds with a fixed, limited depth: roof-tops; over subways; over below-ground facilities; or multistory building set-backs (Figure 13.10). To design an adequate root zone volume, the media unit weight, volume, and water management is crucial. Nutrient management, while important in media design, can be supplemented after installation in a nutrient management program tailored to the plant, media, and irrigation profile. Water demand estimates are influenced by the size and number of plants, heat loading from surrounding structures, wind, and other environmental characteristics of the site. Light patterns will play a role in plant selection.

Media design considerations

Media design and irrigation control are the tools to meet plant water demands. The confines of urban construction in enclosed systems require a designed medium since natural soils are not optimally effective in these

Figure 13.10 Roof-top landscapes re-create a landscape in a thin veneer of soil. Containment and control of irrigation and drainage must be considered to protect the building. (Photo credit: E. Gilman.)

unnatural situations. Increasing soil fines content and organic compost components can increase plant available water, however excessive fines (without soil aggregation) can increase a medium's wet unit weight excessively without increasing **plant available** water holding ability. Excessive organic components will lead to media shrinkage as a consequence of decomposition. References on designed media, such as Phillip Craul's texts[12,38] and the recommendations from U.C. Davis[39] can assist with first approximations for the arboricultural or soils consultant on the design team (Figure 13.11).

Adapted from University of California: U.C. System for Producing Health Container-Grown Plants mix recommendations (1957)

Components:

Sand:
70-85% fine sand (0.5mm - 0.05 mm)
<< 15% passing the #200 sieve
<12-15 % in the coarse sand size class

Organic material:
sphagnum moss peat
rice hulls
bark

Figure 13.11 Basic components of the UC container mix recommendations.

The media should be designed with input from the team's landscape architect, arborist, engineers, and soil scientist. Any sand amendments should also include organic amendments. Lightweight aggregates, despite increased cost, should at least be considered to meet weight and porosity demands in the soil profile. Excessive compost materials result in shrinkage of the media over time, and changes in the hydraulic character of the total container system.

Light-weight soil textural amendments can add porosity and some compaction resistance to a medium without adding excessive weight:

- Perlite-vermiculite-lava: Expensive; many high in fluoride; susceptible to crushing and weathering; can add plant available moisture, air-filled porosity, and nutrient-holding capability. pH: generally it is slightly alkaline for perlite and vermiculite, but variable by source.
- Isolites, zeolites, calcined shales, and exploded, vitrified slates: Expensive; aggregate strength variable; susceptible to weathering; can add plant available moisture, air-filled porosity, and nutrient-holding capability to the plant.
- Pelletized clay products: Expensive, but effective in increasing available moisture, air-filled porosity, and nutrient-holding capability to the plant.

Rooftop or setback planting

Trees and very large shrubs should be supported by load-bearing elements of the underlying structure, but given the ability of a crane to position materials on the roof, the weight of the installation equipment may be the most important consideration in the design of the supporting structure. While more soil is preferred for tree establishment, roof-top situations work on minimums. Irrigation controls can influence media design choices and total volume requirements, since water is the currency of plant survival. Planter weight can fluctuate widely as water is added, drained, or moved through the plants into the atmosphere. The unit weight of the media and its saturated weight should be considered in the design of the planter and the supporting structure. The load must include projected plant weight over time, the saturated weight of the media, peak pedestrian loads, and environmental factors such as snow loads.

Usually the container must be sealed to prevent leakage into spaces below the planter, and drainage must be provided to prevent water logging and plant death. Attention to longevity and detail are crucial since trouble shooting years after the site is completed is extremely problematic, especially if the trees grow to warrant protection resulting from their increased value on the established site. This presents the additional complication for intensive rooftop or setback landscapes since materials need to be staged on the ground rather than on the sealing membrane. Drainage systems need to be designed to minimize holes in the gasket to move water out of the planted site.

Containerized plants and planting beds

In addition to the irrigation, drainage, and nutritional considerations of growing a tree in a confined space, container edges (as much as 6 inches from the edge) are also susceptible to potentially damaging temperature extremes in the root zone, so smaller containers or narrow planting zones could limit root growth due to sidewall effects by temperature and water. Container size should also account for the "sail effect" of the canopy on the potentially restricted root zone to prevent wind-throw of the trees. Also, an adequate root plate zone with additional volume for the critical root zone over time should be provided.

Select tree species which are more tolerant of the extremes of containerized planting in irregular light, wind, or temperature patterns depending on the site. The arborist should be able to supply this type of information.

Estimation of water use and soil volume

As stated earlier, water demand estimates are influenced by the size and number of plants, heat loading from surrounding structures, wind, and other environmental characteristics of the site. Plant water use has been estimated by linking a tree unit water use to meteorological pan evaporation data, and scaling the information to the whole tree. This water requirement can then be imposed on the water holding capacity of the landscape soil during the longest period between rain or irrigation events based on regional soil and climate data.

For many deciduous trees, in most areas of the United States outside the desert south-west, the soil estimation model by Lindsey translates into about 2 cubic feet of soil per square foot of land surface under the tree dripline as a convenient first estimation.[14] Since most of the tree's roots are commonly found in the surface first foot of soil, twice the surface of the envisioned tree's dripline could be projected to meet tree demands without supplemental irrigation. If irrigated during the longest period without rain, the volume could be cut by 50%, or more, depending on the frequency of planned irrigation.

The Lindsey model can be considered conservatively accurate, however Craul has suggested a more comprehensive biophysical approach in contained planting system design. Kopinga's work in Holland suggests consideration into ground-water influences in his region, resulting in a 0.75 soil volume: 1 canopy area ratio, matching the soil nitrogen mineralization/ tree nitrogen requirement budget. The models do not account for root-zone overlap in grove planting.

The same methods can be adapted for media water-holding design and irrigation management planning. The volume should be calculated since rules of thumb cannot eclipse a rational design method given regional differences in rainfall and soil profiles. During establishment of the newly transplanted landscape, daily monitoring of irrigation needs for the first weeks and weekly monitoring for the first year of active growth are necessary to ensure success.

The tree pit as a dysfunctional design

Trees provide shade to pedestrian corridors, malls, and parking lots. The presence of trees positively influence parking behavior, a point not lost on businesses and property owners, who will pay premium prices for the presence of trees. The cooling effects of a shading canopy has been shown to lower gas tank temperatures, thus lowering vaporization losses, adding to the aforementioned environmental benefits of a healthy urban forest.[40] The challenge is the seemingly lack of compatibility of trees and pavement.

The common 4 by 4 foot tree pit detail and the modifications to 4 by 6 foot, for street trees is a dysfunctional design. Tree root systems obviously require greater volumes of soil than are provided in the standard tree pit. Assuming you could even get the root zone to a depth of 3 foot, a 6 by 6 pit is the smallest feasible opening, based on Urban's minimum critical volume of 3.7 cubic yards in a Washington, D.C. climate.[28] Cultivating trees to the ages and sizes which accrue benefits beyond installation and maintenance costs, requires a larger soil volume unless an intensive irrigation and fertilization budget can compensate for the restricted soil volume.

Escape of roots into the surrounding pavement section materials compromises the durability of the pavement. The materials in the pavement section design are chosen for pavement durability, and are compacted to levels considered detrimental to tree growth. The materials in the standard base layer exclude organic materials and minimize fine materials, both of which are necessary for water-holding and nutrient holding capability for tree growth. Roots do escape the tree pit, often following layer interfaces of dissimilar materials or avenues of lesser density (porosity) due to flaws in compaction. Since roots need oxygen, and the two immediate layer interfaces (wearing surface-base and base-subgrade) intersect the root zone, root growth will penetrate. As the layers are relatively shallow, radial expansion of roots can lift pavement causing premature damage and replacement to sidewalks, coupled with legal liability with the tripping hazard.

Tree pit grates are usually detrimental to the established tree. The large metal grates are expensive and require a pro-active maintenance commitment to enlarge the opening prior to the tree trunk cutting into the restricting sides as the tree grows. The heat from the cutting torch will also damage the tree if the operator is not very careful or the tree has grown into the grate material. A mulch covering is highly preferable to a tree grate, but there is the problem of keeping the mulch from migrating out of the pavement opening. ASTM D 448 #3 or #4 aggregates bound with a material such as an asphalt mastic or decomposed rice hulls (Stabilizer®, San Marcos, CA) is a potential solution to this problem. Unit pavers or "Belgian blocks" are two more expensive, but highly attractive options, although there are concerns with providing vandals or rioters with ammunition. The pavers can be placed as concentric circles around the pavement opening and the roots can heave them without excessive damage if they are not removed in a timely manner. To provide air and water movement through the pavers while grouting them to prevent theft, a coarse single-sized sand can be mixed with an asphalt mastic (10 gallons of C.P.E. emulsion per ton of sand) to produce a popcorn grout material.[41]

Tree guards are no more than an aesthetic choice. Cylindrical metal tree guards used to be needed to protect tree trunks from horse bite or tie damage. The horse has been replaced by the bicycle in this scenario. Vandalism by snapping off trees is also a consideration, but the expense of the tree guards compared to the cost of a tree small enough to be snapped would suggest holding money out for replacements and investing it into the maintenance phase of the project. The rest of the money saved in tree guards could be used for bike racks, pedestrian signing, or benches.

Alternative street tree-pavement systems

Several strategies overcome the tree pit issue: Avoidance through creative design is often the best solution. Not planting is an option also. If the trees must be surrounded with pavement, several avoidance methods are available, but are often expensive. Examples of such methods are cantilevered pavements and use of walled–off street cellars to provide large containers. Media to integrate the root zone and pavement do exist. Different approaches can meet loading requirements for the root zone. Most systems attempt to form a skeletal matrix of aggregate, and impose a horticulturally beneficial dimension to the matrix. The risk is the disruption of matrix formation as the horticultural amendments are added. The type of wearing surface is also important, since unit pavers on pedestrian areas might allow some small settlement whereas a concrete sidewalk cannot escape breakage given the same loss of support. It is up to the client to decide, as tolerance for unit paver settlements might be fine for "historic feel" projects, whereas you might get sued by a student who trips on the ice-covered lip of a sidewalk crack.

Cantilevered pavements

Cantilevered pavements are expensive, but they are commonly observed to be a feasible method of constructing pedestrian surfaces over tree root zones. Two points are worth mentioning. Compaction control is essential. Very expensive applications have been destroyed when installing contractors compact the entire rooting zone to peak density (AASHTO T-99) operating on standard construction procedure. The root zone must not be compacted. The materials supporting the edges of the suspended concrete slabs of metal framing for concrete support must be of a quality material and compacted. If the purpose for the cantilevered concrete is to encourage tree survival and sustained growth, then the opening must be large enough to accommodate trunk caliper growth. I have seen 4 inch diameter openings in sectioned concrete for 3-3.5 inch caliper tree transplants, which will not work if the trees are expected to grow at all.

The planting soil under pavement will usually be a specified material replacing the existing materials due to previous compaction and activity on the site. It should not be compacted at all. Local soil specialists should be consulted to meet regional soil availability and regional standards. A coarse sandy loam with organic compost amendments should work if the materials are available.

Unless there is naturally occurring deep drainage out of the planting zone, positive drainage and a passive aeration system similar to the container system should be designed. Irrigation of the system can be part of the passive system, or access tubes through the pavement such as the W.A.N.E.® tree system (W.A.N.E. Tree Systems; Tampa, FL) could be employed.

The size of the cantilevered zone is variable. The shared tree rooting volume, running as a wide tree planting strip along the street provides volume and the advantage to connected rooting volumes in drainage planning. An alternative is to have large isolated zones, like a "macro tree pit." In either geometry, there needs to be enough width to accommodate the trees' root plates in the mature design.

Combination container-cantilever systems

Many cities have basement entrances to older commercial buildings or cellars extending under the pedestrian corridor. Innovative public works personnel, such as D. Gamestetter in Cincinnati, have used those cellars as large bottomless containers after walling–off the building access for tree establishment.[42] In Cincinnati's case, they have naturally sandy subsoils providing drainage out of the containerized system. Containerized systems extending partially above and below–grade can gain depth of soil volume when surface-areas are restricted by other structures.

Unless an urban landscape design revolution occurs, or municipalities yield unlimited funding to tree planting projects, there will be trees in pedestrian corridors partially or totally surrounded by paved surfaces. Since cantilevered pavements are expensive, and not always an appropriate solution, methods to integrate root zones and pavement sections should be explored. Many of the comments for cantilevered sidewalks in reference to dimensioning, maintenance, and drainage apply for this approach. In cases of reactive soils or freezing climates, the best strategy would be to avoid vertical interfaces of dissimilar materials under the pavement surface to prevent differential movements and cracking.

Amsterdam tree soil

Mixture: 90% by volume coarse angular sand: 10% fine organic compost (Full documentation and description can be found in Couenburg, 1994.[43])

The material is well suited for unit paver construction with limited loading (pedestrians only) and assumes a high factor of internal drainage below the planting zone. The material cannot be compacted to standard AASHTO levels without limiting root penetration into the material.[44] The evolution of the material design accepted 1 inch settlements after paving with unit pavers. The suggested level of compaction is gauged to a penetrometer resistance of 1.5–2.0 MPa (217–290 psi) rather than a relative density measure.[43] Estimates of relative compactness range from 70–80% peak standard proctor density. The penetrometer reading is taken with an agricultural field penetrometer rather than an AASHTO hammer penetrometer. The material can work within these confines and support very acceptable tree growth. Given the potential for settlements, its use is probably limited to unit pavers. The volumes proposed for the material are based on root plate development, regional water availability, or irrigation regime.

CU Soil®

Mixture: Available from local licensed distributors
78–85% DOT approved crushed aggregate by weight
15–22% soil
0.0234–0.0255% acrylamide hydrogel

The material was developed as a relatively low-cost sub-base pavement material for tree establishment.[45] The mix design process produces a material which produces a California Bearing Ratio greater than 50 when compacted to peak density (AASHTO T-99). The material assumes drainage provided out of the planting area. It is intended to be used under the entire pedestrian mall, sidewalk, or parking lot without compromising pavement wearing surface durability. It can be used as isolated large planting zones or planting strips under pavement receiving vehicular traffic. The base material should be USCS GP preferably angular with high void ratio after full compaction. The CU Soil® method produces a gap-graded material. The mixture's stone produces a load-bearing skeleton with viable soil suspended within the skeleton voids. For this reason, a crushed stone is preferred to provide the highest void content to maximize soil before a loss of strength. The hydrogel is used to stabilize the mixture for mixing uniformity, transportation, placement, and vibratory compaction.

CU Soil® is a relatively new media design. Bench-scale testing of the material began in 1993. The first reduction to practice occurred in 1994, and the patent was issued in 1998. Long-term tree evaluation is not available. Based on the first three year controlled paved field study at Cornell University, tree establishment was comparable to an agricultural field soil, and superior to conventional sidewalk installations.[46] While not observed to be limited in the establishment phase, irrigation and nutrient management systems are likely to be needed over time. The extent of necessary irrigation and nutritional needs and strategies for management are still in the research phase. Uncontrolled working applications in New York City and Ithaca, NY display acceptable tree establishment and growth. Several hundred trees have been installed into close approximations of the material during its development and have not been critically evaluated. Over two hundred trees are the subject of long-term observation in New York, but reliable data is not available. Survival percentages have been high and establishment growth has been acceptable for the species observed.[45]

Designing solutions

The following sections illustrate examples of tree establishment techniques through drawings and photographs.

Berms or raised planters

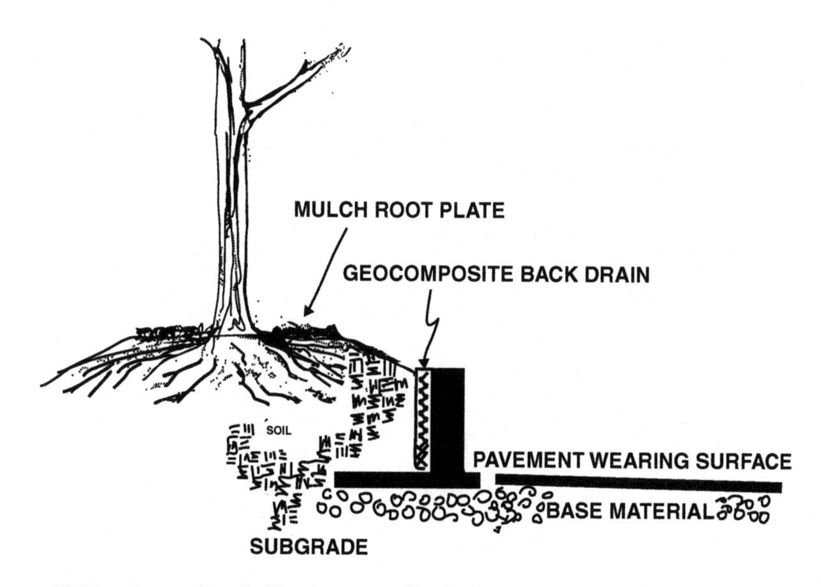

Figure 13.12 Berm detail. The berm wall, if used, is protected by a geo-composite strip drain product as a back drain for the wall and for the berm. A geo-textile may be used for aggregate separation at the soil/pavement base material interface. The top of the berm wall may be used for pedestrian seating.

I. Berms are used to raise the planting over an unsuitable subgrade, augment a root volume depth over a suitable subgrade, provide a visual barrier, or provide a noise barrier.
II. Berm width should equal tree root plate: Minimally 1 meter for small trees, 2 meters for large. The length will be used to supply soil volume.
III. Top of berm should be level for water infiltration, with the root plate area mulched (2 inches maximum).
IV. Sides should be sloped at a low angle (3 length :1 rise) to prevent erosion, or walled to contain the sides.
V. Drainage should be greater than one inch per hour or else supplemented by under-drains.
VI. Irrigation nozzles should be micro-emitters specified to each plant to prevent runoff and increase efficiency.
VII. The top of the wall can be a seating area or pedestrian control.
VIII. Walls discourage pedestrian traffic over the root systems.
IX. Low shrubs or ground covers may be planted on the berm, but the root plate area will benefit from lack of competitive planting, keep the root plate area weed free and mulched.
X. A sandy loam soil could be used for the berm, amended with a low level (< 5% by volume) of organic compost. Consult with your soils expert for regionally standard and available materials.

Figure 13.13 Berm planting on the University of Florida Campus parking lot. Note the width and sloped sides to the pedestrian walkway. (Photo credit: J. Grabosky.)

Figure 13.14 The same berm is capped at the end with walls to direct pedestrian traffic (Brick in background).

Vegetation strips and inter-connected tree pits

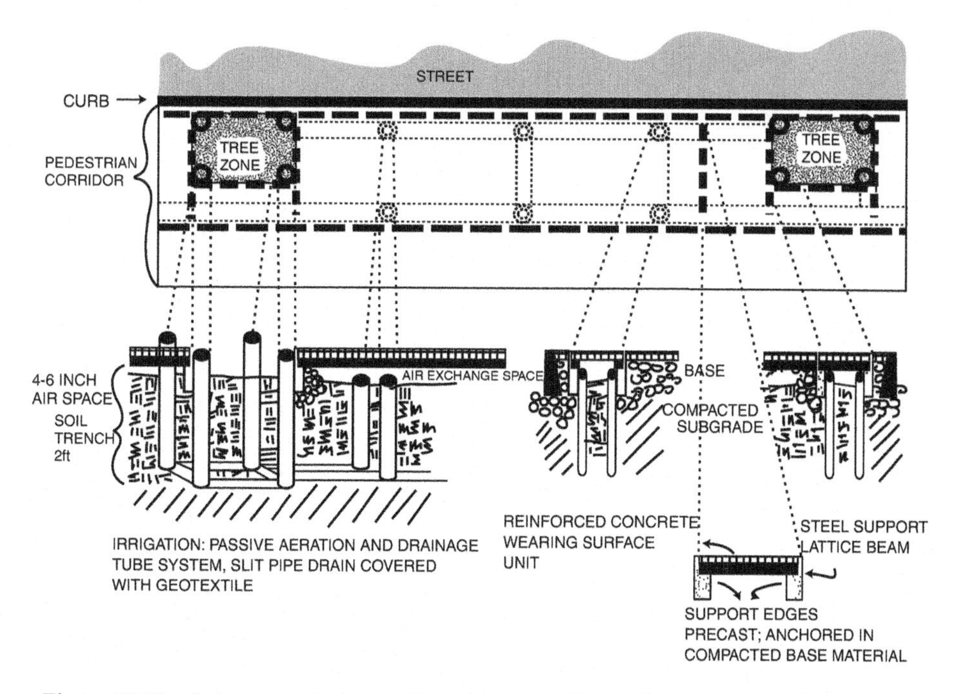

Figure 13.15 Interconnected tree pits using a cantilevered pavement with a passive aeration-manual irrigation system.

Figure 13.16 A vegetation strip needs to be large enough for a root plate (6 feet) and provide a viable soil material for plant growth. The soil needs to be either protected throughout construction, which is difficult, or replaced with a specified soil material. If the soil is replaced over a compacted subgrade, or if natural drainage is limited, then designed drainage is necessary to prevent periodic flooding. (Photo credit: J. Grabosky.)

Rooftops and setbacks

Figure 13.17
Figure 13.18a
Figure 13.18b
Figure 13.19

Figure 13.17 Shared rooting zones along an access road at University of Florida at Gainesville. The root zones are disconnected for pedestrian access to buildings. Irrigation has been provided, compensating for the relatively small root zones allowed by the necessary buildings and access corridors. (Photo credit: J. Grabosky.)

Figure 13.18a Examples of possible uses for conduits to expand the root zone under structural limitations. The figures also point out that given the right soil situation (type and parent material), some species can exist near pavement without necessarily causing damage. (Photo credit: J. Grabosky.)

Figure 13.18b Continued from Figure 13.18a. (Photo credit: E. Gilman.)

Containerized planting

I. Stationary containers are very similar to berms when connected to soil (bottomless) and roof-top/building set-backs (sealed bottom).
II. Stationary containers are limited to larger containers in northern climates due to temperature influences, but benefit from access to designed irrigation systems.
III. Portable containers should be designed for movement such as pallet fork access.
IV. Temperature influences should be considered for extreme heat and cold situations. Near-surface soil volumes should be subtracted from containers exposed to extreme sun or winter cold.
V. Volume influences irrigation frequency and ultimate plant size.
VI. If drainage out of the container onto the surface treatment is allowed, the drainage water may stain the surface near the container.

Protection of trees during construction

Protection of trees during the construction process applies the same principles of soil and water protection and management used in designing plant establishment zones. The plants must be evaluated by a consulting arborist to determine if the plant material is worth saving. The evaluation requires

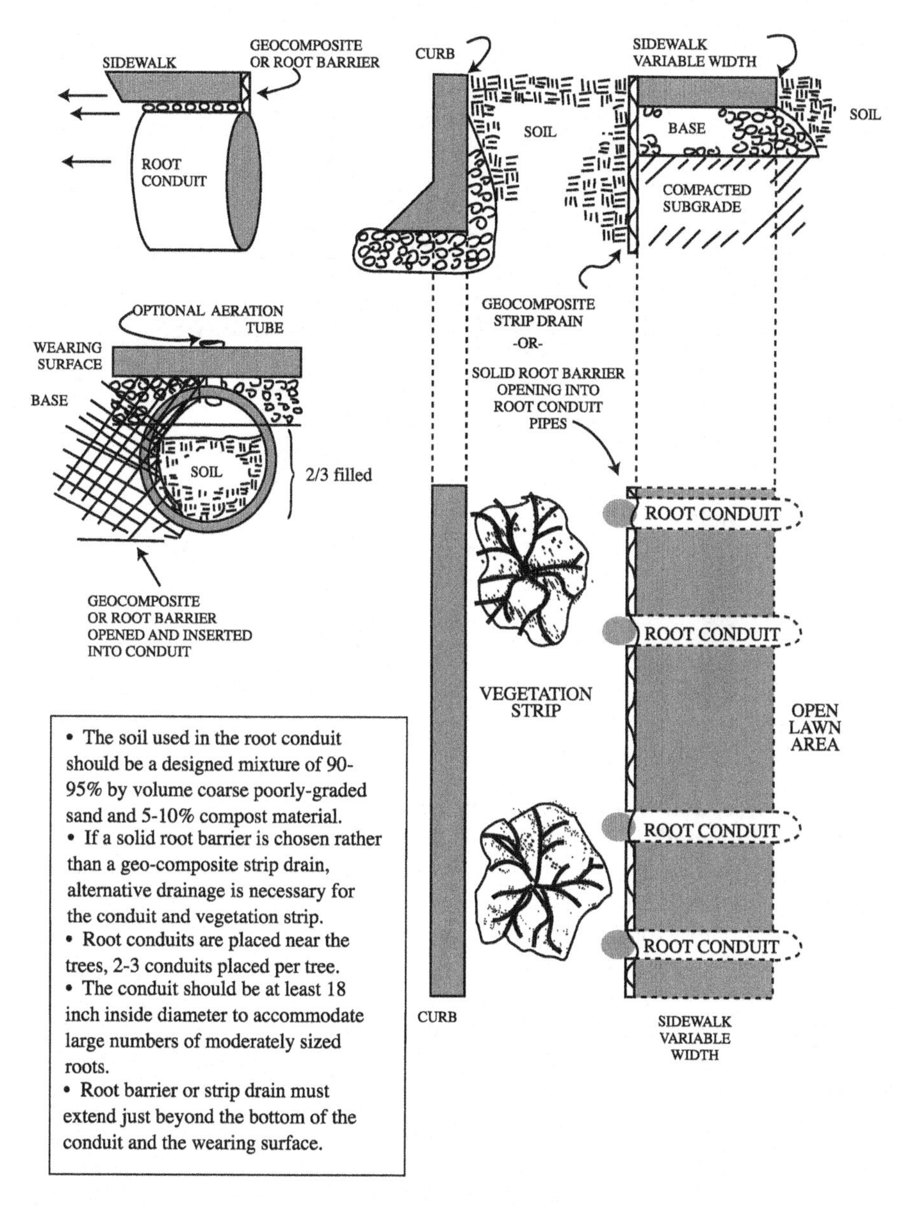

Figure 13.19 Details of the untested conduit system to bridge root zones under paved surfaces.

knowledge of site conditions, specific tree information, predicted tree response to the construction event, and costs of tree care during and after construction. Particular information for the tree includes: species, age, condition, potential hazard, historical significance, and aesthetic value of the

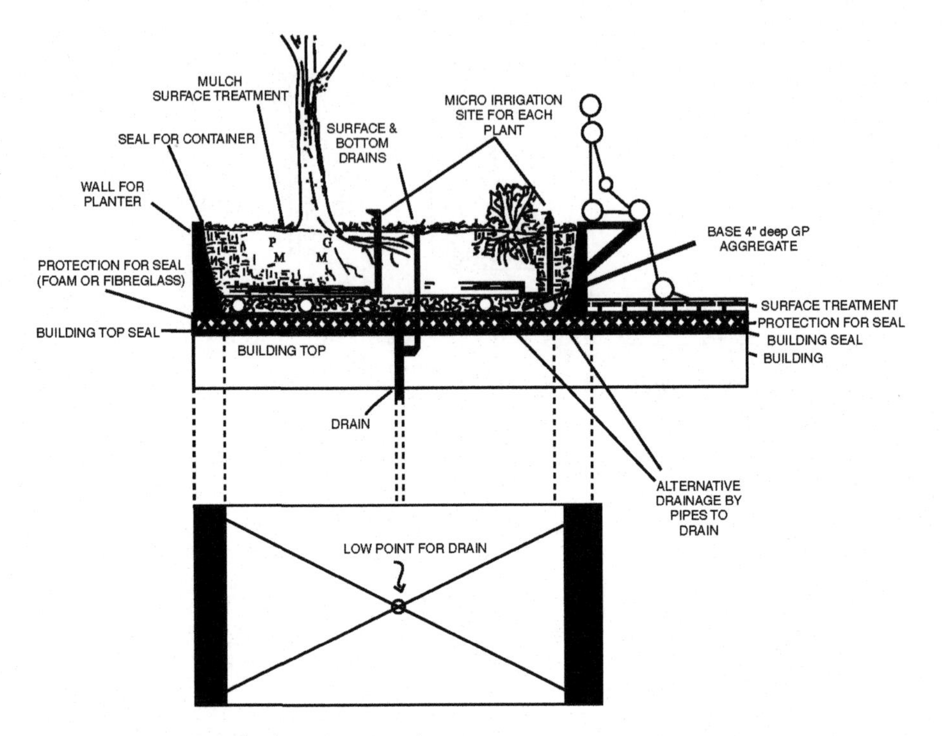

Figure 13.20 Detail of roof-top application.

individual tree, and collective canopy.[47] Every situation is unique, and once the tree is damaged, it only seals off the injury. Trees never heal, actually they can be used as historic record. Trees are dynamic organisms, growing in response to their surrounding environment. Rapid changes in soil volume, wind direction and speed, light regimes, and surrounding plant competition will influence plant response to the stresses of construction activity, and structural stability of the tree after construction activity has ended.

Once a mature tree is removed, it cannot be easily replaced with an equivalent plant, so thought should go into the decision to cut. If the change to the site is thought to cause the tree to become an unacceptable risk or hazard, it should be removed. Removing over 50% of the root system or necessary trenching to the trunk of the tree on any side could fit this category. The tree may survive in decline for several years, shedding large branches, but a fast death and replanting may be preferable to the public opinion polls during the long, slow death. The replacement tree could be providing shade while the damaged tree is soaking up tree maintenance and removal budgets. If over 20% of the canopy must be removed, call for an arborist's opinion. A tree may have aesthetic value, but reliable power, water, and sewer systems need to be the priority. If the tree is of great value, then structures and utilities are certainly less expensively retooled than the heroics to work through the tree without damage.

Figure 13.21 Interior drainage geocomposite for a sealed container positioned on the sides of the container. (Photo credit: E. Gilman.)

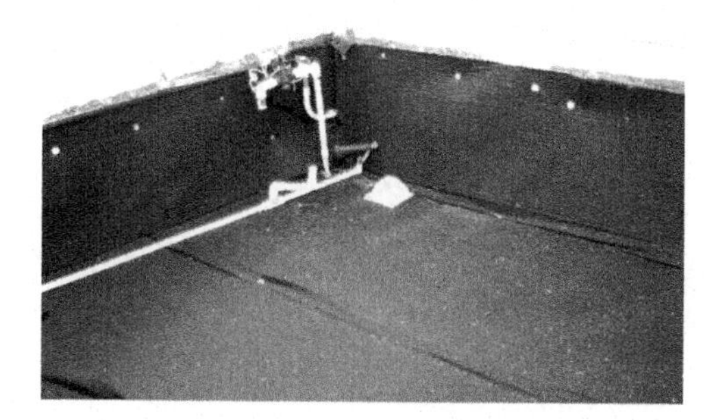

Figure 13.22 Example of irrigation access for a sealed container. (Photo credit: E. Gilman.)

Designating the protection zone

Protection of the root system is crucial for tree survival and recovery during construction. Assignment of the protection zone is based on the current size of the tree or the envisioned size of the tree in the mature design. Soil excavation is necessary to determine the geometry of the root mass. This is

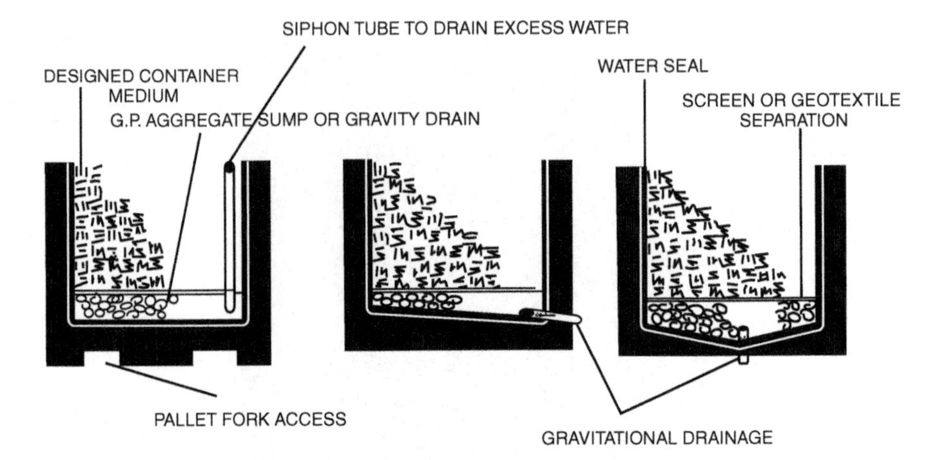

Figure 13.23 Container detail examples.

Figure 13.24 Stationary container with 2 irrigation risers. Drainage is gravitational out to the pavement surface treatment. Note the discoloration near the container. (Photo credit: J. Grabosky.)

accomplished manually with shovels or soil probes, or mechanically with compressed air jets, working inward toward the trunk from some distance away from the tree. The flares at the trunk are indicators of major roots radiating from the trunk, but roots do not follow a set geometry. Root growth is influenced by plant competition, water, nutrient supply, wind direction, mechanical loading from the trunk, and penetrability of soil.

Figure 13.25 Containerized trees in raised planters. The size of the containment should match the size of the plant material at its ultimate size. (Photo credit: E. Gilman.)

There is general agreement that trees can lose substantial percentages of their root mass and survive. Younger trees can lose major portions of their root system and survive, (balled and burlapped trees often loose 90% of their roots in harvesting)[48] while older trees are much less tolerant of such events. The stress of root loss is aggravated by other construction factors such as increased heat loading, increased water demand, and loss of soil volumes for root regeneration due to compaction or structures. Providing irrigation, professional pruning, and attention to nutrient management before, during, and after the construction phase can minimize impacts of development on large trees.

Recent models of critical root protection zones suggest preservation of at least 60% of the total root mass and total protection of the "root plate." The percentage of total root mass is related to resource demands, the root plate is related to stability under wind or snow loading. Since the root system is not visible, several estimation methods are used to calculate a critical root zone. For the Dripline methods, multiply the dripline by 1.5 to 2 for area estimation. The method does not work well for narrow trees, often found in urban designs. In those cases, 40% of the tree height has been suggested. For the Trunk formulae, multiply the trunk diameter (in inches) by 1.5–2.5 for a protection zone diameter in feet. Once this "critical' zone is established, it should accommodate future growth by a factor of 2.25.[49,16]

The basic architecture of most mature tree root systems consists of five or fewer major structural roots radiating from the trunk: all the secondary or lateral roots; further growth to the fibrous "feeder" roots; to the white, fleshy absorbing roots; and one-cell root hairs ultimately connect to the "plumbing" of those few major roots to move materials into the canopy of the tree. Removal of a major root, or several roots greater than one inch, can eliminate large percentages of the total root system. Those root losses can severely damage portions of the canopy and compromise stability of the tree. Tree response to damage can occur immediately or over a period of decline up to five or more years. Delineation of the protection zone must

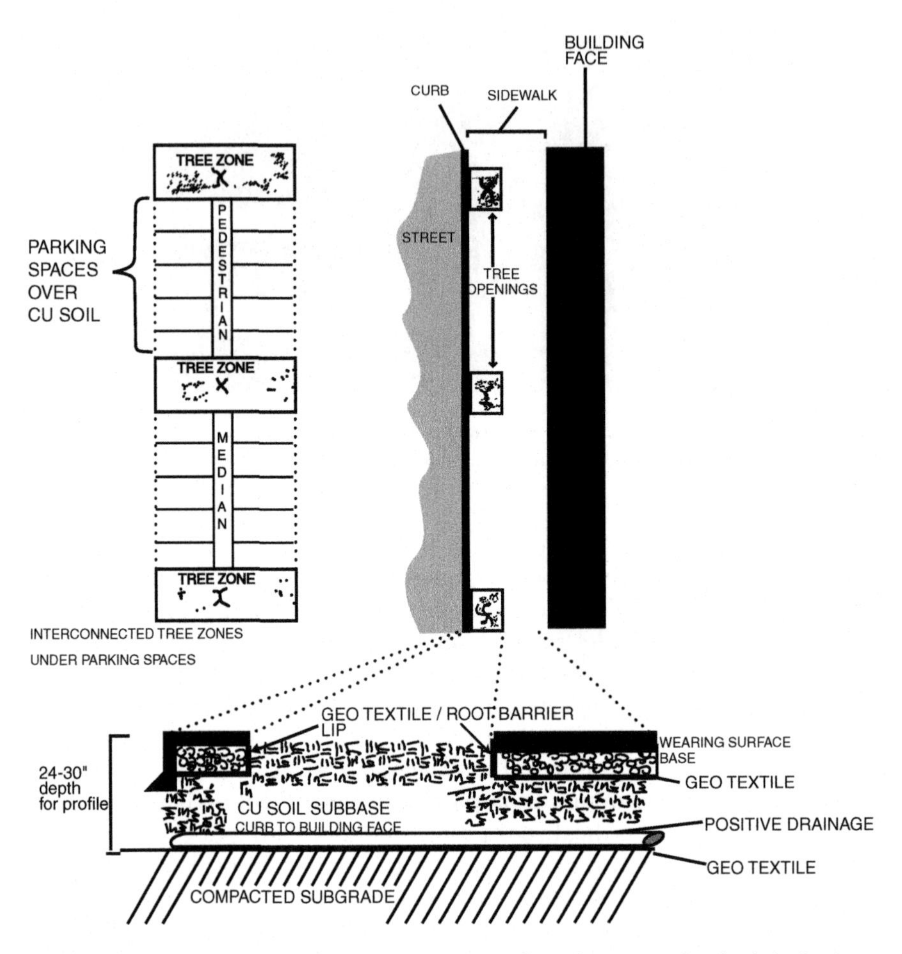

Figure 13.26 Examples of a possible skeletal soil applications. On the left: for interconnected parking lot root zones. On the right: when there is no room for a vegetation strip between the street and the building, but a requirement for emergency or maintenance traffic on the pedestrian corridor.

consider the root plate and the position of the finer root zone of colonization. Conservation of viable soil volumes is more cost effective than ameliorating or compensating for a compromised system. Protection zones must anticipate future growth whenever possible.

Roots do not have to be cut to sustain damage; compaction or fill materials also have detrimental impacts on tree roots. Soil compaction can be broken into three groups: purposeful compaction for construction needs; compaction as a cost of the business of construction on the site; unnecessary compaction due to poor planning or poor monitoring on the site. Given the value of healthy trees for a developed property, the increased expense in working around or with trees in the design and construction phases of a project is compensated by a safer canopy, and increased value of the final

Figure 13.27 CU Soil test profile showing three years of root growth. Note the downward growth habit of the crabapple root system. This profile was paved as a sidewalk for field testing of the material in early research testing.

project, or client relationship. Staging of materials off site, equipment and contractor parking zones, just-in-time deliveries, controlled machinery access to future roads, or staging materials in the future parking areas can minimize the environmental "cost of doing business."

Strict protection language with specified replacement penalties in the construction contract, and their enforcement, must be in place, assuming a specified value is assigned for each key plant or zone, and protection areas are clearly defined. An extreme example has been to post the value of each tree on the plant, and charging full replacement value even for relatively minor cosmetic damage, the policy being stated in the contractual agreement. More often the penalty values are assigned based on a damage appraisal by a consulting arborist, which could be modified to include damage amelioration and replacement of several smaller trees for each specimen lost or damaged. The penalty should be severe to prevent damage incurred when the perceived labor savings is greater than the preservation costs: Those "tree value" decisions are made in the design phase, not in the construction phase.

Protection zone delineation should be made by solid structures such as fences with clear markings. Silt fencing and orange plastic fences may work, but are easily breached by equipment and materials. Several opinions exist for the determination of the protection zone. Certainly the larger the protection/preservation zone is superior to a restrictive zone, but a method of estimation should exist. Field wisdom suggests using the edge of the canopy, or dripline as a first estimation. There are potential problems in this method if the tree is to be encircled by development since roots are known to grow well beyond the dripline of many trees common in urban planting. The critical root zone by Coder, or the soil volume method of Lindsey can be used for estimating. All the estimations contain assumptions that are subject to some interpretation; and certainly trees function well in soil volumes much smaller than the models would suggest. The estimations are subject to information on root colonization, local soil characteristics, water availability, and construction practicality. The estimates serve as a benchmark in assigning initial plant protection zones that are ultimately subject to site specific negotiations. Ultimately, frequency of watering events and level of care will determine volume estimation.

If a critical zone for planting must be crossed with equipment, the soil could be protected by stretching and securing a geotextile over the zone and using a deep wood mulch over the textile. The textile must be secured to use its load distributing properties as it stretches, or else the system may not be effective, and the textile only assists with the clean-up operation. If there is a major route for equipment over a zone, steel beams supporting timber could be used to limit the compaction to the beam zones, that could be tilled after removal. Take care not to destroy the area you are protecting in the cleanup operation. This requires hand labor in mulch removals.

References

1. McPherson, E.G., D. Nowak, and R. Rowan, Eds., *Chicago's Urban Forest Ecosystem: Results of the Chicago Urban Forest Climate Project*, USDA Forest Service N.E. General Tech. Report NE-186, (Radnor, 1994).
2. Taha, H., H. Akbari, and A. Rosenfeld, Heat island and oasis effects of vegetative canopies: micro-meteorological fields measurements, *Theor. Appl. Climatol.*, 44:123–128, 1991.
3. Bassuk, N. and T. Whitlow, Evaluating Street Tree Microclimates in New York City, in: L.G. Kuhns and J.C. Patterson, Eds., *METRIA 5: Selecting and Preparing Sites for Urban Trees* (U.S. Forest Service, NE Area, 1985), 18–27.
4. Dochinger, L.S., Improving City Air Quality with Trees, in: *Forestry Issues in Urban America: Proceedings the Society of American Foresters Regional Conference*, New York City, pp.113–120, 1974.
5. Kristofferson, P., Climbing Plants on Walls – Advantages and Disadvantages, in: G. Watson and D. Neeley, Eds., *Trees and Building Sites: Proceedings of an International Conference Held in the Interest of Developing a Scientific Basis for Managing Trees in Proximity to Buildings*, Morton Arboretum: International Society of Arboriculture, Savoy, Illinois, pp. 88–98, 1995.
6. Huang, J.H., H.T. Akbari, and A. Rosenfeld, The potential of vegetation in reducing summer cooling loads in residential buildings, *J. Clim. Appl. Meteorol.*, 26:1103–1106, 1987.
7. McPherson, E.G., L.P. Herrington, and G.M. Heisler, Impacts of vegetation on residential heating and cooling, *Energy Build.*, 12:41–51, 1988.

8. Dwyer, J., The Significance of Trees and Their Management in Built Environments, in: G. Watson and D. Neeley, Eds., *Trees and Building Sites: Proceedings of an International Conference Held in the Interest of Developing a Scientific Basis for Managing Trees in Proximity to Buildings*. Morton Arboretum: International Society of Arboriculture, Savoy, Illinois, pp. 3–11, 1995.
9. Cook, D.I., Trees, Solid Barriers, and Combinations: Alternatives for noise control, in: *Proceedings of the National Urban Forestry Conference; Syracuse, NY.* SUNY ESF Pub. 80-003. pp. 330–339, 1980.
10. DeWalle, D.R., Amenities Provided by Urban Plants, in: D.F. Karnosky and S.L. Karnosky, Eds., *Improving the quality of urban life with plants: Proceedings of the June 21-23 1983 International Symposium on Urban Horticulture*, Bronx, NY. The New York Botanical Garden Institute of Urban Horticulture, 7–14, 1983.
11. Heisler, G.M., How Trees Modify Metropolitan Climate and Noise, in: *Forestry Issues in Urban America: Proceedings the Society of American Foresters Regional Conference*, New York City, pp. 103–112.
12. Craul, P.J., *Urban Soil in Landscape Design*, John Wiley & Sons, Inc., New York, 1992.
13. Kopinga, J., Evaporation and Water Requirements of Amenity Trees with Regard to the Construction of a Planting Site, in: D. Neeley and G. Watson, Eds., *The Landscape Below Ground II: Proceedings on an Internation Workshop on Tree Root Development in Urban Soils*, San Francisco, CA: International Society of Arboriculture, Savoy, Illinois, pp. 233–245, 1998.
14. Lindsey, P. and N. Bassuk, Redesigning the Urban Forest from the Ground Below: A New Approach to Specifying Adequate Soil Volumes for Street Trees, *Arboric. J.*, 16:25–39, 1992.
15. Harris, J.R. Irrigation of Newly Planted Street Trees, in: D. Neeley and G. Watson, Eds., *The Landscape Below Ground II: Proceedings on an International Workshop on Tree Root Development in Urban Soils,* San Francisco, CA: International Society of Arboriculture, Savoy, Illinois, pp. 225–232, 1998.
16. Harris, R.W., J. Clark, and N. Metheny, *Arboriculture: Integrated Management of Landscape Trees, Shrubs, and Vines*. Prentice-Hall, New Jersey, 1999.
17. Kielbaso, J.J. and V. Cotrone, The State of the Urban Forest, *Make Our Cities Safe for Trees: Proceedings of the Fourth Urban Forestry Conference,* St. Louis, MO, National Urban Forest Council, pp. 11–18 1989.
18. American Forests, The State of Our Urban Forest: Assessing Tree Cover and Developing Goals, *American Forests*, 1997.
19. Pokorny, J.D., *Urban Forest Health Needs Assessment Survey: Results and recommendations*, USDA Forest Service, N.E. State and Private Forestry: St. Paul, MN, NA-TP-01-98, 28 p., 1998.
20. Moll, G. The State of Our Urban Forest, *American Forests*, November/December 1989, 61–64, 1989,
21. Kopinga, J., Aspects of Damage to Asphalt Road Pavings Caused by Tree Roots, in: G. Watson and D. Neeley, Eds., *The Landscape Below Ground: Proceedings on an International Workshop on Tree Root Development in Urban Soils.* Morton Arboretum: International Society of Arboriculture, Savoy, Illinois, pp. 165–178, 1994.

22. McPherson, E.G. and P. Peper, Infrastructure Repair Costs Associated With Street Trees in 15 Cities, in: G. Watson and D. Neeley, Eds., *Trees and Building Sites: Proceedings of an International Conference Held in the Interest of Developing a Scientific Basis for Managing Trees in Proximity to Buildings.* Morton Arboretum: International Society of Arboriculture. Savoy, Illinois. pp. 49–63, 1995.
23. Mudrick, S.P., Sidewalk Displacement: Fullerton Grinds Out A Solution, *Public Works*, August 1990, 50–51, 1990.
24. Rolf, K., O. Stal, and H. Schroeder, Tree Roots and Sewer Systems, in: G. Watson and D. Neeley, Eds., *Trees and Building Sites: Proceedings of an International Conference Held in the Interest of Developing a Scientific Basis for Managing Trees in Proximity to Buildings.* Morton Arboretum: International Society of Arboriculture, Savoy, Illinois, pp. 68–77, 1995.
25. Stal, O. and K. Rolf, Tree Roots and Infrastructure, in: D. Neeley, and G. Watson, Eds., *The Landscape Below Ground II: Proceedings on an International Workshop on Tree Root Development in Urban Soils.* San Francisco, CA: International Society of Arboriculture, Savoy, Illinois, pp. 125–131, 1998.
26. O'Callaghan, D. and M. Lawson, The Potential for Foundation Damage Caused by Tree Roots, in: G. Watson and D. Neeley, Eds., *Trees and Building Sites: Proceedings of an International Conference Held in the Interest of Developing a Scientific Basis for Managing Trees in Proximity to Buildings.* Morton Arboretum: International Society of Arboriculture, Savoy, Illinois, 99–107, 1995.
27. Cutler, Interactions Between Buildings and Roots, in: G. Watson and D. Neeley, Eds., *Trees and Building Sites: Proceedings of an International Conference Held in the Interest of Developing a Scientific Basis for Managing Trees in Proximity to Buildings.* Morton Arboretum: International Society of Arboriculture, Savoy, Illinois, 78–87, 1995.
28. Urban, J., New Techniques in Urban Tree Plantings, *J. Arboric,* 15(11):281–284, 1989.
29. Arnold, H.F., *Trees in Urban Design*, Van Nostrand Rheinhold, NY, 168 p., 1980.
30. Kopinga, J., Research on Street Tree Planting Practices in the Netherlands, *Proc. 5th Annual METRIA Conference*, Pennsylvania State University, University Park, PA, 1985.
31. Perry, T.O., Conditions for Plant Growth, in: P. Rodbell, Ed., *Make our Cities Safe for Trees: Proc. of the fourth urban forestry conference;* St. Louis, MO, Oct. 15-19, 1989, Washington, D.C., American Forestry Association, pp.103–110, 1990.
32. Moll, G. and J. Urban, Giving Trees Room to Grow, *Am. Forests*, May/June: 61–64, 1989.
33. Bakker, J.W., Growing Site and Water Supply of Street Trees, *Green*, 39(6):205–207, 1983.
34. Mattheck, C. and H. Breloer, The Body Language of Trees: A handbook for failure analysis, *Research for Amenity Trees No. 4.* 4th impression, The Stationary Office, London, 1998.
35. British Standards Institution, *Guide to Trees in Relation to Construction, BS 5837*, BSI, Linford Wood, Milton Keynes, MK14 6LE, 1991.
36. Bloniarz, D.V. and Ryan, H.D.P. III, Designing alternatives to avoid street tree conflicts, *J. Arboricul.*, 19(3):152–156, 1993.
37. Grabosky, J., *Street Tree Rootzones* (lecture), American Society of Consulting Arborists 29th Annual Conference: Oct. 9–12, Seattle WA, 1996.
38. Craul, P.J. and J. Lienhart, *Urban Soils: applications and practices*, John Wiley & Sons, NY, 1999.

39. Baker, K. F., *The U.C. System for Producing Healthy Container-Grown Plants, Agr. Exp. Sta.* Manual 23, 1957.
40. Scott, K.I., J. Simpson, and E.G. McPherson, Effects of tree cover on parking lot microclimate and vehicle emissions, *J. Arboricul.*, 25(3):129–142, 1999.
41. Evans, M.D., *Concrete Block Paving and Its Effectiveness as a Pavement Around Street Trees*, MLA Thesis, Cornell University, Ithaca, NY, 1989.
42. Gamstetter, D., Designing the right tree for the right place, *Arborist News*, June 1998, 9–12, 1998.
43. Couenburg, E., Amsterdam Tree Soil, in: D. Neeley and G. Watson, Eds., *The Landscape Below Ground: Proceedings on an International Workshop on Tree Root Development in Urban Soils.* Morton Arboretum: International Society of Arboriculture, Savoy, Illinois, pp. 24–33, 1994.
44. Couenburg, E., *Amsterdam Tree Soil: penetrometer case studies*, Field presentation at the 1999 International Society of Arboriculture 75th Annual International Conference, Stamford, CN. August 1–4, 1999.
45. Grabosky, J., N. Bassuk, L. Irwin, and H. van Es., Pilot Study of Structural Soil Materials in Pavement Profiles, in: D. Neeley, Ed., *The Landscape Below Ground II: Proceedings of an International Workshop on Tree Root Development in Urban Soils*, San Francisco, CA, International Society of Arboriculture, 210–221, 1999.
46. Grabosky, J., N. Bassuk, L. Irwin, and H. van Es, Structural Soil Investigations at Cornell University, in: D. Neeley and G. Watson, Eds., *The Landscape Below Ground II: Proceedings of an International Workshop on Tree Root Development in Urban Soils*, San Francisco, CA, International Society of Arboriculture, 203–209, 1999.
47. Barrell, J., Pre-development Tree Assessment, in: D. Neeley, Ed., *Trees and Building Sites: Proceedings of an International Conference Held in the Interest of Developing a Scientific Basis for Managing Trees in Proximity to Buildings.* Morton Arboretum: International Society of Arboriculture, Savoy, Illinois, 132–142, 1995.
48. Watson, G., Root Development After Transplanting, in: G. Watson and D. Neely, Eds., *The Landscape Below Ground: Proceedings on an International Workshop on Tree Root Development in Urban Soils.* Morton Arboretum: International Society of Arboriculture, Savoy, Illinois, 54–68, 1994.
49. Coder, K., Root Growth Control: Managing Perceptions and Realities, in: D. Neeley and G. Watson, Eds., *The Landscape Below Ground II: Proceedings of an International Workshop on Tree Root Development in Urban Soils*, San Francisco, CA, International Society of Arboriculture, pp. 51–81.

chapter fourteen

Building for high performance

Hillary Brown

Contents

Overview 194
Putting the forces to work 196
'Beneficial occupancy' redefined 197
  Seeing the light 197
Breathing easier — indoor air quality (IAQ) and healthy inhabitants 199
  Contractor's stake in IAQ 199
Reuse it or lose it — waste prevention and waste management 201
  Choose to reuse — the build less approach 201
  Other waste prevention and management strategies 202
  Recycled content and the use of renewable resources 203
From the ground up and the top down — realigning the construction process 205
Materials, means, and methods revisited — the contractor's role 205
  Explicit specifications 206
  Commissioning 206
  A blueprint for change 206
Settling accounts 206
The (hidden) high costs of construction 207
May the force be with you! — energy efficiency examined 209
  Siting and passive solar design 210
  The building envelope 210
  Daylight and daylight controls 211
  Lighting and equipment 211
  Load management, right sizing, and controls 211
"Design integration" 214

0-8493-7486-3/01/$0.00+$.50

The road to renewables 214
Good housekeeping 215
Brighten your "black-holes" 216
Gimme shelter? — besting infestations of pests! 217
Green n' clean 217
References 217

Overview

Many attentive developers harvesting savings from energy efficiencies have also leveraged some compelling additional benefits. Building occupants have responded to more extensive daylighting, use of quality, high-efficiency lighting, and improved indoor air quality with an increased sense of well-being. This may translate into improved levels of performance in the workplace. As such, these productivity savings could dwarf combined capital, operations, and maintenance cost savings of the building.[19]

As we learn from these high performance achievements, we should be inspired to improve the performance levels of *all our buildings*, embracing emerging technologies — from today's off-the-shelf efficiency measures to tomorrow's collectors of energy income from the sun, the wind, and the earth. To do so, will however require new visions for how we build, with increased sensitivity to human and environmental outcomes. It will involve new architectural and engineering collaborations, more involved operating and maintenance efforts, and perhaps also new mechanisms for how we finance building.

In the next millennium, global energy, and environmental concerns as well as health issues will increasingly dominate building economics. The construction industry will be adjusting to these imperatives. Today's builders of "high performance buildings" are ahead of the curve. They are proactively committing to maximizing operational energy savings, providing optimized health interiors, and limiting the detrimental impacts of the construction on the environment. And as a result, these developers and building owners are reaping meaningful cost savings and important benefits associated with energy and resource-efficient facilities (Figure 14.1).

Conventional building projects generate material waste, energy inefficiencies, and pollution emissions, all true costs not showing up on any balance sheets. Economists call them "externalized costs," meaning the environment in particular, and eventually society in general will absorb them. Smart developers have begun to invert these liabilities into economic opportunities by adopting new technologies and materials to realize both environmentally sound *and* profitable projects. Smart construction industry professionals will both understand the issues and know the means and methods of constructing "smart" as it becomes more important and economically beneficial to owners. As the industry becomes motivated through the perception that "green" is a source of competitive advantage, it will engage energy and resource concerns

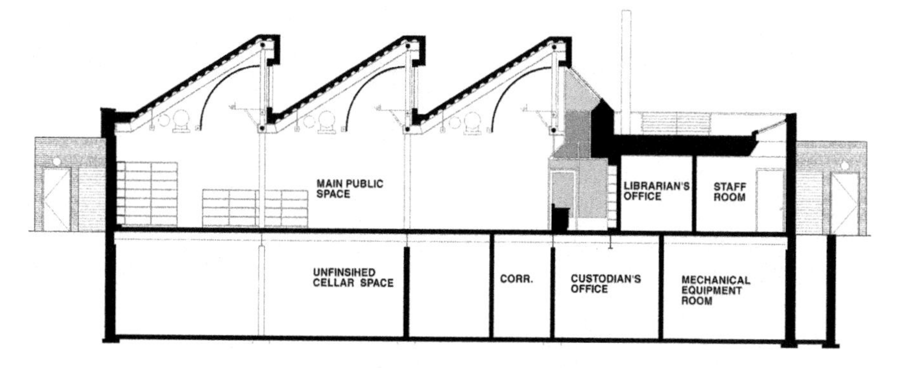

Figure 14.1 Given the siting constraints of this new branch library, the roof is the primary envelope element available as an interface with the natural environment. The south-facing monitors introduce sunlight for direct heat gain during the winter, and lighting year-round. During the cooling seasons, automated shades limit the light to just the levels needed for library functions. The peaks in the roof collect the hotter air, which during the winter is circulated through the building by the HVAC system. During the cooling season, this hot air is exhausted. Curved diffusing baffles and reflective light shelves prevent direct sunlight from reaching the occupied areas of the building. The light fixtures are controlled by photo-sensors, filling in whatever portion of the required levels are unmet by daylight. South Jamaica Branch Library. (Illustration: Stein White Architects, LLC.) Queens Borough Public Library, New York City Department of Design and Construction. (Illustration: Stein White Architects, LLC.)

as a catalyst to create fundamentally better buildings. Higher targets for environmental performance will allow private enterprise the latitude to innovate and deliver. Eventually, public demand will move high performance over the top of the curve and into conventional practice.

Worldwide, the real estate and building industry has, for some time, taken notice of its own shortcomings, recognizing the magnitude of its contribution to problems threatening global economics, health, and well-being. In Europe and Japan, regulations and building codes are fostering widespread use of daylighting and renewable energy technologies. For many years, Sweden has subsidized loans for an energy-efficient residential sector. Mexico has imposed water consumption limits for plumbing fixtures operating in its water-poor capital city.[1] But the distance to go is considerable. In the developing world, which is experiencing an unprecedented construction growth spurt, western patterns of land use and building design are replacing climate-specific, vernacular building types which traditionally have had fewer environmental impacts.

In this country, we are just seeing the beginning of an industry-wide shift in attitudes towards environmental stewardship in building. Federal, state and municipal policy makers across the country have begun to capitalize on the economic and environmental benefits of green buildings built for their own portfolio of facilities. A large number of U.S. Department of

Energy and Environmental Protection Agency programs, in particular, the Energy Star Building Program, have a track record of encouraging partnerships with the commercial sector to move the real estate market towards high performance.

The insurance community is also looking at various high performance technologies as a long-term means of achieving traditional risk management objectives. It perceives a link between poor indoor air quality and increased health insurance and professional liability claims. As risk managers, the industry foresees its sector's increasing exposure from storm damage, flooding, and other property loss predicted as a likely outcome of global climate change. Insurers therefore are looking at the loss-prevention benefits of energy-efficient and renewable technologies.

As new commercial real estate practices emerge in response to consumer awareness of green building benefits, there will be a growing need for an objective process for measuring and advancing these goals. The U.S. Green Building Council (USGBC) is an example of a non-profit consensus-based membership organization representing all segments of the building industry which, through its mission to accelerate green building practices, is responding to these perceived needs. USGBC has developed a national green building rating system for commercial buildings, called LEED or Leadership in Energy and Environmental Design. LEED represents a long-term program to unify practices through educational tools provided through a common rating and certification system.

Big savings in fuel and electrical energy costs, together with significant environmental benefits, can be achieved by integrating today's "off-the-shelf" technology into both new construction and renovation. Lockheed's Building 157 in Sunnyvale, California brought extensive daylighting into the building through 15 foot high window walls and a skylit atrium, saving 75% of its lighting bill or approximately $500,000 annually in energy savings. The City of San Diego renovated a conventional 1980's suburban office by upgrading mechanical, lighting, and electrical systems, yielding an annual savings of approximately $76,000 or 60% less than the operating costs of the original structure.[2] A showcase remodeling of offices for the Natural Resources Defense Council in New York reduced lighting energy consumption 75% and cooling and heating consumption 50% and 75%, respectively.[3] This owner went on to tally the potential contribution of rolling out similar retrofits in all buildings nationwide over 20 years; it would mean elimination of 175 millions tons of carbon dioxide emissions and saving $60 to $80 billion in avoided costs of power plant construction.[4]

Putting the forces to work

The city of San Diego produced an exemplary integrated design approach to energy efficiency. It employed simple "state-of-the-shelf" technology in its renovation of the Ridgehaven Office building, one of a pair of identical typical 1980's suburban offices. It used Department of Energy DOE 2 computerized energy modeling to predict future operating savings, justifying the retrofit of solar film to windows, additional shading, and added insulation to improve the envelope. It replaced inefficient mechanical systems with high efficiency water source heat pumps, adjustable speed condenser pump drives, variable air volume boxes for outside air, and DDC controls. Designers reduced the lighting usage using 40% fewer lamps with electronic dimming ballasts, and daylight sensors and occupancy sensors controlling the lighting levels. Compared to its un-upgraded twin, Ridgehaven now saves approximately $76,000 in annual energy bills or 60% of its original operating costs.[4]

The recently completed SC Johnson Wax's new world headquarters in Racine, Wisconsin offers a number of artful approaches to energy efficiency. The design looked at how and when people occupy spaces and designed systems to respond to those active or sedentary needs. The common areas — circulation zones along the exterior windows and the atrium — were treated as spaces occupied *transitionally,* spaces where one could tolerate higher summer temperatures, and lower winter temperatures. The workstations, however, offer individualized comfort controls, bringing air through a raised floor plenum. This project, which incorporated a lot of daylighting, also used computer energy modeling programs and large-scale physical models to predict cooling, heating, and electrical loads. The scheme used air-to-air heat exchangers to recover heat exhausted from the building's laboratories. Predictions are for energy savings of 60% over a comparable facility.[5]

'Beneficial occupancy' redefined

Beneficial occupancy is the term commonly used in construction to signify the stage when the owner may assume the intended use of the facility. As we examine some of the qualitative improvements promised by high performance design, this term assumes new meanings. The term "beneficial occupancy" here may apply to the much-enhanced visual, thermal, and air quality features of a high performance building.

Seeing the light

Daylight is a basic driver of human actions; daily and seasonally, its varying intensity paces our lives and affects our moods. As we spend over 90% of our time inside, it is important that we invoke it indoors. Properly played off a building's walls, ceilings, and intermediate surfaces, it provides higher quality illumination than electric light. More spectrally rich than artificial illumination (with more visible wavelengths), it raises visual acuity. Its dynamics provide visual stimulation and information about the outdoors. Together, these attributes are believed to affect occupant well-being and performance.

The beneficial effect of daylight in learning areas was an organizing principle in the design of several Johnston County, North Carolina public elementary and middle schools. The premise was that daylighting could contribute to the mental and physical well-being of students and teachers, while at the same time lowering overall life-cycle facility costs. Large roof monitors were placed above the major occupied spaces, the classrooms, gym, cafeteria, and media center. The monitors were equipped where necessary with low "E"(low emissivity) glazing or light baffles to eliminate glare. Light level sensors in ceilings determined adequacy of daylight energy in the space and adjusted conventional fluorescent lighting. Smaller windows added view but did not significantly contribute to the day light effect. The overall design effect achieved illumination of 70 plus foot candles during two-thirds of the occupied time. On the energy side, the results were impressive: the additional cost investment for the daylighting paid for itself in operating cost savings in under three years, after contributing to a 66% reduction of electrical lighting loads compared to the non-daylit schools. Balancing heating, lighting, and cooling needs meant an average 60% reduction in overall energy costs.[6]

What was even more compelling were the performance benefits measured by the architects, who analyzed data from the students' achievement tests.[7] County-wide comparisons were made with other students who learned in new schools but non-daylit environments. The tests compared relative increase in performance test scoring within the various schools over a five year period, looking at end-of-grade, average reading, math, and the standard achievement tests. The students who attended the daylit schools out-performed their peers in new non-daylit schools. The improved performance ranged from 5% for one year to an average of 14% for multiple years. In the same test, the same age group of students housed in temporary mobile classrooms (no natural lighting) were evaluated, and artificial lighting actually showed up *as a negative impact* on student performance, with an overall decrease of 17% in performance.

In 1992, a study in Alberta, Canada replicated results from an earlier (1981 to 1985) study, which examined the *qualitative* impacts of different lighting systems. Full spectrum fluorescent lighting is the system most closely resembling natural daylight, containing low-level ultra-violet bands. Far from having neutral effects, the full-spectrum fluorescent lighting actually promoted several performance benefits. Attendance increased in these schools, by an additional 3.2 to 3.8 days per year. Full-spectrum lighting induced significantly more positive moods in students than conventional cool-white lighting, and affected some activity levels and behaviors (e.g., lower noise levels in the libraries.) Because of the trace amounts of ultraviolet light which helps children synthesize vitamin D, students exposed to this full spectrum lighting had significantly less dental decay and grew in height an average of 2.1 cm more over the two year period of the study.[8] These two case studies together are interesting examples which support the potential link between qualitative aspects of light and occupant performance.

In a 1993 paper entitled *Greening the Bottom Line – Increasing Productivity through Energy-Efficient Design*, the authors set out compelling economic justification for the creation of high performance workplaces. One example cited was a post office in Reno, Nevada where a $300,000 investment in an energy efficiency retrofit provided qualitative improvements in lighting through use of daylight and improved artificial lighting. These changes precipitated real improvements in output. A six percent increase in the rate at which workers sorted mail, combined with a reduction in errors, generated productivity gains worth $400,000 a year, dwarfing the energy savings, and reducing the payback for the upgrade to just nine months.[9]

Breathing easier — indoor air quality (IAQ) and healthy inhabitants

It seems obvious to all that building occupants should enjoy good quality indoor air which is free of pollutants. Conventional building practices, however, may no longer guarantee that basic entitlement. According to the federal Environmental Protection Agency, unhealthy indoor air may plague more than 30% of new or renovated buildings world-wide.[10] What is referred to as sick building syndrome (SBS) encompasses a wide range of short-term complaints such as headache, fatigue, difficulty concentrating, respiratory tract irritations, and infections. In order to avoid the potential of being involved in a costly legal battle, contractors need to protect air passageways during construction, and balance and commission the start-up of the systems to achieve optimal operation.

Contractor's stake in IAQ

Many common building materials contain volatile organic chemical compounds (VOCs) which are emitted or "off-gassed" as they cure during construction and on into occupancy. Without adequate ventilation, the undiluted presence of these compounds may cause temporary discomfort, although long-term exposure may entail more serious health risks. Finishes, cabinetry, and furniture typically emit VOCs such as benzene, formaldehyde, styrene, chemicals often used as binders, adhesives, sealants or solvents in paints. Once released, these compounds may migrate within the building and become temporarily adsorbed in other material 'sinks' such as carpet or fabric, only to be re-released by them later on. *Microbial* organic compounds may be formed from mold and bacteria in the presence of heat and moisture. As the building ages, some binders may deteriorate or experience erosion by air currents, releasing further VOCs.[10] Housecleaning chemicals are another continuous source of VOCs (Figure 14.2).

The contractor participates in administering an IAQ management plan during construction. This is vital for obtaining the value-added features of good indoor air quality. Sequencing the work is also important. Scheduling the installation of all wet material (sealants, adhesive, paints) well *before* installation of absorbent dry materials (carpet, tile, fabric panels) can prevent the latter from becoming "sinks" for chemicals off-gassed during wet material curing, and later releasing them into occupied spaces. The contractor

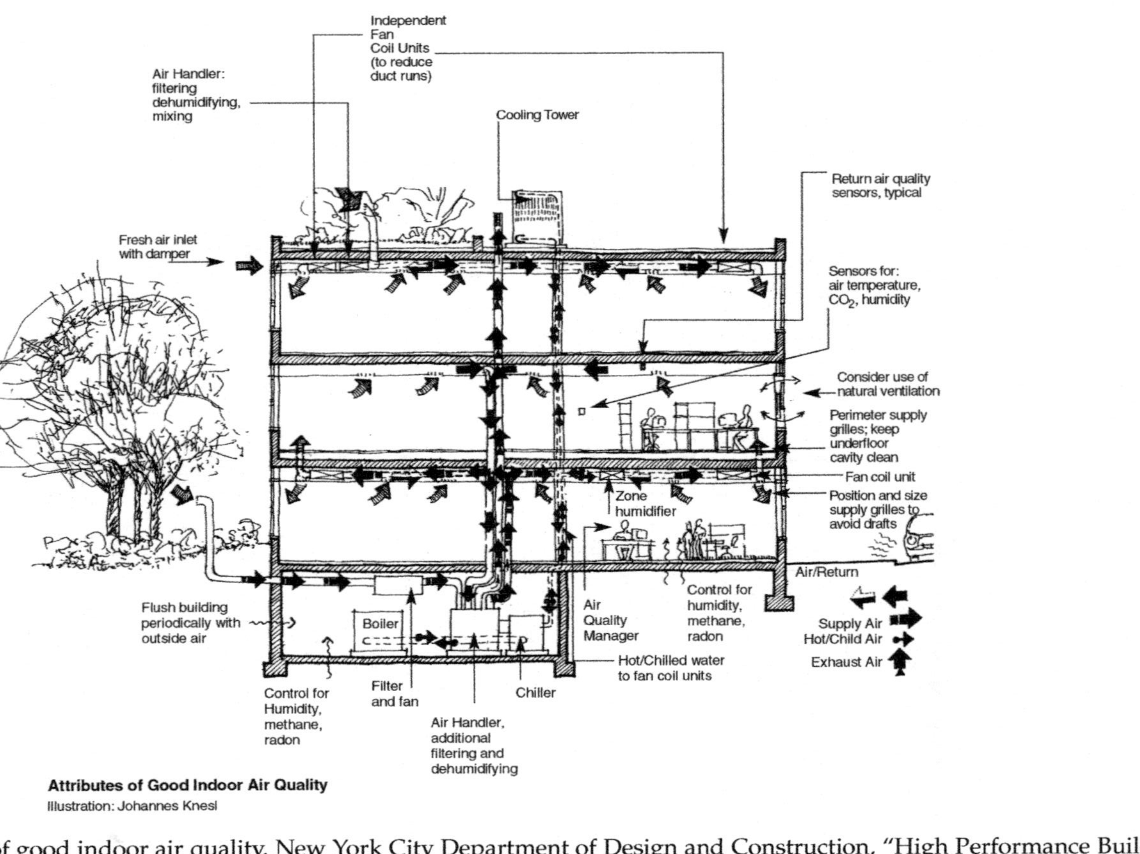

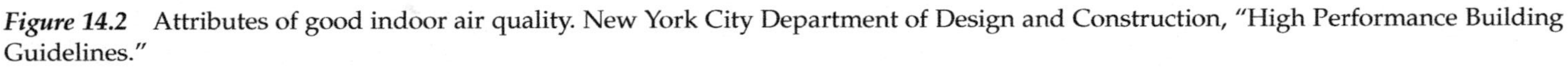

Figure 14.2 Attributes of good indoor air quality. New York City Department of Design and Construction, "High Performance Building Guidelines."

needs to coordinate the actions of subcontractors to ensure this proper sequencing.

During construction, controlling moisture through adequate ventilation, or if required, inducing positive pressurization to limit outside moist air infiltration, are measures to prevent microbial growth in insulation or on finish materials. Contractors need to also keep ductwork free from dust, debris, and moisture. Prior to occupancy, a good practice is to allow a building *flush-out period* in which tempered 100% outside air is allowed to carry off remaining levels of VOCs. Duct, air shaft, and general area cleaning with high efficiency particulate air (HEPA) vacuums should be a final measure before occupancy. Though construction is short in duration compared to the life of the building, the lasting environmental benefits can be obtained only through conscientious construction administration.

Reuse it or lose it — waste prevention and waste management

Every year, the building industries extract more than 3 billion tons of raw material from the earth's crust and transform it into building fabric. Quarried stone, sand, and ore become concrete, brick, and steel. We incorporate nearly a billion cubic meters of wood annually into construction.[12] These hidden costs are reflected in the value of construction. We can employ a more savvy use of resources during routine construction transactions. Resource management during construction involves accounting activities which help make us aware of waste, and motivate the construction team to curb or avoid it, beginning with whole building *waste prevention* approaches.

Choose to reuse — the build less approach

High performance projects begin with an evaluation of "no-build" or "build *less*" options. These could include adaptive reuse of an existing structure as a realistic alternative to building new, or innovative space planning resulting in less need for new square footage. As technology transforms our workplaces and institutions today, it is no longer axiomatic that every program function deserves a dedicated spatial solution. *Telecommuting* (the use of home as office), *universal sizing* (standardizing workstation sizes for function interchangability), and *alternative officing* (creating unassigned, flexible workstations to be shared by multiple users) are effectively downsizing facility construction in many organizations.

Even though preservation and adaptive reuse of existing older structures has long been common practice, few analyses make clear the economic and environmental advantages of building reuse vs. *new* construction. These would calculate, for example, the *avoided cost* of demolition, removal and site clearance (necessary for most new urban construction), which can be significant. Clearly, there is also *economic value* inherent in the mass of an older structure, in the high quality of its craftsmanship, in its embodied energy — the energy gone into mining, fabricating, and installing materials and finishes. The generous floor to ceiling heights enable older buildings to more readily accept the retrofit of today's power, HVAC, lighting, and telecommunications requirements. Oversized window walls admit maximum light and fresh air. Both features are often costly to replicate today. For its relocated national headquarters, the Audubon Society renovated a landmark-quality commercial structure; this investment retained 300 tons of steel, 1,000 tons of masonry, and 560 tons of concrete. Audubon estimated that reuse of the building avoided what would have been premium costs to demolish and build new, driving the price of their headquarters higher by a third.[13] The industry will incorporate these costs into their plan when it is mandated by the public.

In another example, Southern California Gas Company salvaged major portions of an ordinary 1950s commercial building for their showcase Energy Resource Center in Downey, California. Sections of this office complex were partially dismantled, elements salvaged, and a major new architectural element inserted. A conventional approach — demolition of the structure and building new — was priced at $ 9.7 million. Partial reuse of the building structure and fabric brought the project in at $ 6.7 million. The owner pocketed $ 3 million in savings from avoided demolition, removal, and site clearing costs, and from selected reuse of building components. They also diverted 350 tons of material from the waste stream.

Savings may also result from reusing individual building components or obtaining salvage value for recyclable materials. Any structure to be demolished is a potential source of useful building materials. Windows, doors, slate tile, brick, hardware, and structural lumber can often be refurbished for an extended life. A notable demonstration effort was undertaken by the Ridgehaven Green Office Building which recycled a third of its existing acoustical ceiling tile, most doors, frames, and hardware and rebuilt its existing office furniture system and interior partitions. For offsite salvage, the building yielded reusable carpet, light fixtures, and heat pumps. The project recycled 28 tons of metal construction debris, 7 tons of wood scrap, and over 4 tons of cardboard and packaging material,[14] and it was cost-effective to do so.

Other waste prevention and management strategies

Careful consideration of waste prevention early in the project can yield considerable resource economy. Accurate purchasing by contractors can avoid accumulation of unused material much of which often ends up in the waste stream. Quality assurance measures prevent the wastage of construction mistakes, where work must be torn out and redone. The careful protection of stored materials can also check waste due to damage or loss. Other prevention measures may involve distributor redemption of unused materials or the returning of packaging such as wood palettes, plastics, and cardboard for reuse, which can dramatically reduce landfill waste.[15]

Demolition and construction waste, once generated at the site, may be best managed by on-site resource recovery as opposed to wholesale disposal. Waste management plans introduce site separation and storing of recyclable materials for future transport to waste processing markets. With landfill closures and tightened regulations on inter-state shipment of municipal solid waste, even the additional labor costs incurred in sorting and transporting waste to recycling centers may be soon be offset by escalating waste tipping fees. Such measures to recover material, together with increased consumer demand for products with recycled content, will stimulate increased profitability in markets for recycled steel and other metals, masonry, sheetrock, lumber, carpet, and tile.

In construction of the new Rose Garden arena in Portland, Oregon, 92% of the waste generated in demolishing the existing site structures was reused or recycled, for a savings of approximately $200,000. A C&D consulting firm identified materials for reuse and recycling, working with sub-contractors on implementation. Almost 45,000 tons of materials were salvaged for reuse on the project.[16]

Recycled content and the use of renewable resources

The green building movement has created a market for new materials — products which in their extraction, manufacture, and installation both minimize consumption of natural resources and avoid detrimental environmental releases. Comparably priced finish products such as carpet, underlayment, ceramic tile and ceiling tile, wallboard, insulation, flooring, fabric, etc., now incorporate a high percentage of post–consumer recycled waste. Steel framing and structural studs are available fabricated with 100% post consumer steel and aluminum. Many products utilize material wastes recovered from industrial processes, for example, fly ash from power plants is used as low-cost additive for concrete (Figure 14.3).

The construction of Duracell International Inc.'s new headquarters in rural Bethel, Connecticut went a long way to reflect the corporation's interest in making its products environmentally friendly. A green team set about maximizing recycled material in the new structure. Almost half of all the materials going into the building and site were selected as containing some measure of recycled content. Examples included the superstructure steel, the fly-ash in the concrete, aluminum roof panels and wall panels, floor tile, and ceiling tile. The building's handsome masonry exterior was comprised entirely of brick incorporating scrap manganese dioxide power, a waste product from Duracell's battery plants.[17]

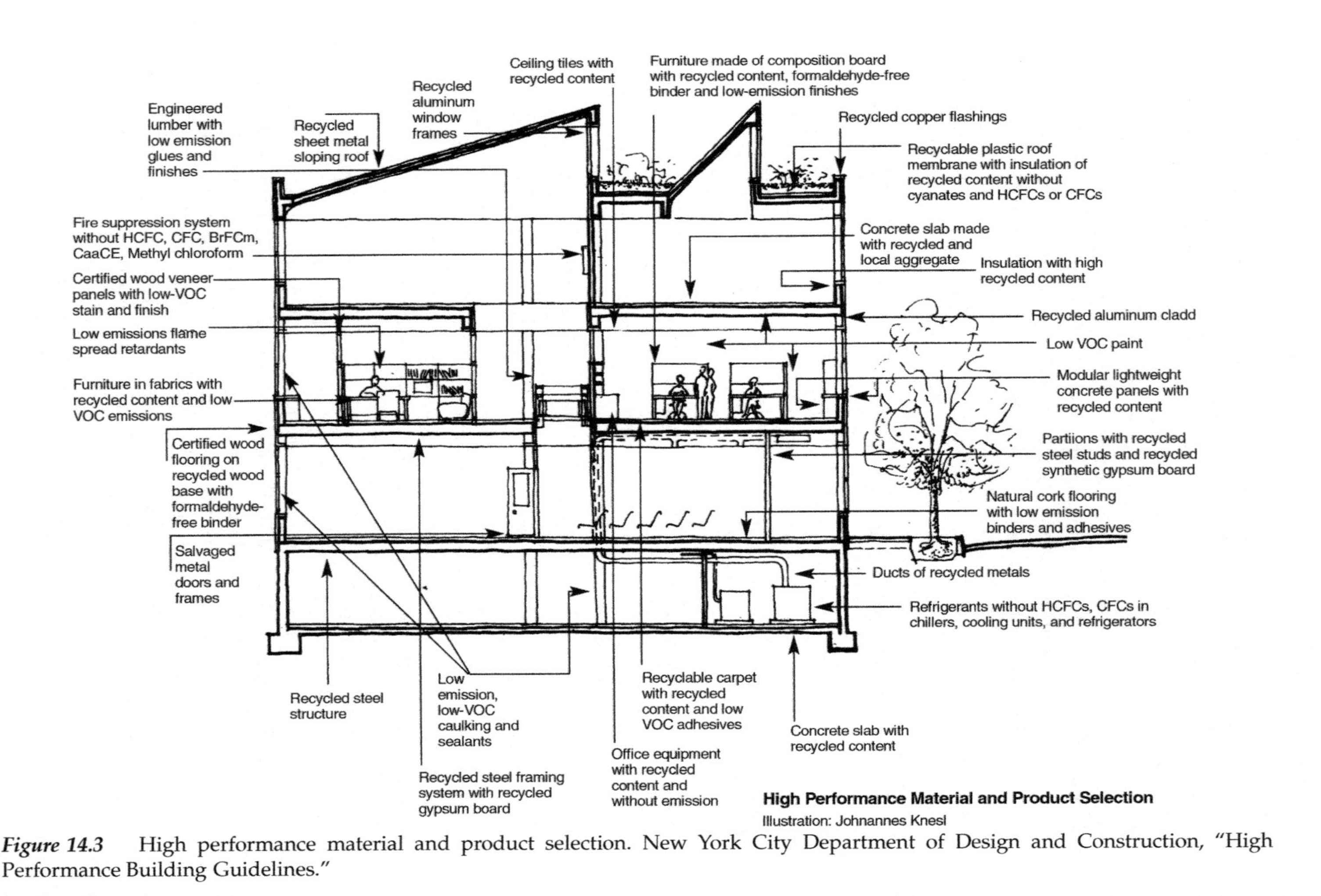

Figure 14.3 High performance material and product selection. New York City Department of Design and Construction, "High Performance Building Guidelines."

From the ground up and the top down — realigning the construction process

High performance outcomes demand a much more integrated approach to the design and construction process and mark a departure from traditional professional service delivery where emerging designs are handed sequentially from architect to engineer to consultant to contractor. A team-driven approach is in fact a "front-loading" of expertise, bringing owner, design professionals, operators, construction managers, and often trade contractors to the early goal-setting process. Cross-disciplinary collaboration is therefore encouraged. In fact, some green developers will begin the building process with one or more facilitated workshops involving all the players in a brainstorming session, breaking down traditional adversarial roles.

Experienced practitioners have emphasized that the collaborative effort and getting all the right players into the front-end of the process is integral to the high performance end product. Input from builders, users, and operators early and often can accelerate the process, eliminate redundant effort by building commitment to decisions, reduce errors, and identify synergistic opportunities. In this way, design/construction/operation strategies forge a seamless continuum.

Materials, means, and methods revisited — the contractor's role

The act of construction by nature invites environmental chaos. It brings on topographic upheaval, generates waste, noise, high impact traffic, and air pollution. Under constraints of schedule and budget, we tolerate this turmoil, despite its stress for existing occupants and neighbors. Many high performance construction activities, however, can reduce these stresses while improving the long term environmental performance of the building.

The contractor should prepare a site protection plan that strategically *reduces* construction impact on the site, and specifies how the site will be utilized for staging, storing materials, moving equipment, controlling stormwater, and providing ventilation. Such a plan also includes approaches for waste management.

To ensure contractor allegiance to environmental goals, getting the contractors and trades to be part of the *up-front* process brings the best results. On the S.C. Johnson building, having the construction team on board during design helped them realize that their activities would require departures from traditional means and methods. As Hellmuth Obata and Kesselbaum senior designer Bill Odell described it: "these contractors sat through months of meetings when there was little for them to do, but this ended up being crucial to our success because they heard all the environmental talk and became sympathetic and involved with what we were trying to accomplish."[18]

Explicit specifications

Project success may also involve obligating the contractors to some alternative means and methods contractually, for example, the need to install "wet" materials before "dry." Environmental goals, unless they are cost neutral, are unlikely to happen spontaneously. Specifications also need to indicate departures from common practice (the flush-out periods, for example) either as stand alone sections or integrated into CSI format. Most green projects include additional language under CSI Division One, a dedicated "environmental procedures section" on general site management. The specifications must also be clearly written to indicate product requirements in terms of environmental performance and make substitutions difficult when an alternatively proposed product lacks similar environmental attributes.

Commissioning

The commissioning process defines additional roles played by project team participants to give an owner assurance that the various equipment and systems and controls providing lighting, heating, cooling, and ventilation in the building work together effectively in conformance with design intent. For best results, the commissioning process should track and complement each of the major milestones of the project development, from programming, design, construction, acceptance, to the post-occupancy phase of the building.

A blueprint for change

A creative owner's or developer's strong stewardship is required to overcome what often appears to be industry indifference, but is better characterized simply as inertia. Leadership in the different sectors that comprise the industry, collaborating on improved standards, incentives, and rewards will go a long way towards inventing a future in which high performance building design, the blueprint for long-term sustainable development, is universally practiced.

Settling accounts

Few people in the building trades realize how buildings directly and indirectly contribute to environmental and human health problems with linked economic consequences. Accounting methods fail to consider the economic relationship of the building's interior environment to health and productivity. Our general accounting practices are likely to undervalue energy efficient technologies or healthy building products by measuring that product or system performance on a first cost basis only, rather than over the building's life cycle. In addition to keeping in place systems of subsidies which fail to reflect true costs, practices also discount the real value of irreplaceable resources as they are expended. Buildings everyday squander valuable capital by wasting energy, water, natural resources, and human labor. Most of them do so inadvertently, by following accustomed practices — often just by meeting (and failing to exceed) building codes.

The (hidden) high costs of construction

The hidden costs of constructing buildings are considerable. They are attributable mostly to the adverse environmental side effects arising from construction-related activity. Today's design decisions have local, regional and global consequences tomorrow. The energy intense production of steel, for example, generates pollution. The mining of ore destroys habitat. As concrete, ordinary brick or tile are produced, the extraction of raw material creates soil disturbance, runoff, and siltation of surface waters. Not only do the product prices not reflect the social cost of these associated damages, they are often artificially reduced by subsidies. For example, it is estimated that in 1996, the federal government spent around $21 billion in taxpayer financed tax credits and subsidies of fossil fuels.[20] It also picks up the clean-up costs from private mining activities on public land and pays for road and infrastructure upkeep for the timber industry.

According to the Worldwatch Institute, almost 40% of the 7.5 billion tons of raw materials annually extracted from the earth get transformed into the concrete, steel, sheetrock, glass, rubber, etc., — elements of our building fabric. Landscape destruction, deforestation, soil and air pollution, toxic runoff are left behind, and 25% of our annual wood harvest is used for construction, contributing to flooding, deforestation, and loss of biodiversity.[21] No accounting system today is capable of accurately pricing this damage to our landscapes — the true depreciation in environmental quality — except perhaps to merely tally lost revenue from agricultural or recreational use.

Product "life-cycle analysis" tries to account for overall economic and environmental impacts by assessing all the resource inputs and waste outputs across a product's life span — from its manufacture, to its use, and its ultimate disposal. Other metrics evaluate the 'embodied energy' or total energy consumed by raw material extraction, transportation, and fabrication *on the way* to becoming a useful product. Embodied energy is usually directly related to the amount of pollution produced. The manufacturing of concrete, glass, steel, aluminum, for example, involve increasing quantities of energy input per pound so their embodied energy becomes a measure of their negative environmental impact (Figure 14.4).

The on-going use and operation of a building exacts a long-term toll on the environment. Buildings utilize about 16% of global water withdrawals: in the U.S., about 55 gallons per person per day. Buildings consume, all told, about 40 percent of the world's energy production, and this energy generation also accounts for the production of about 40% of the sulfur dioxide and nitrogen oxides which cause acid rain and ozone depletion.[22] Of graver

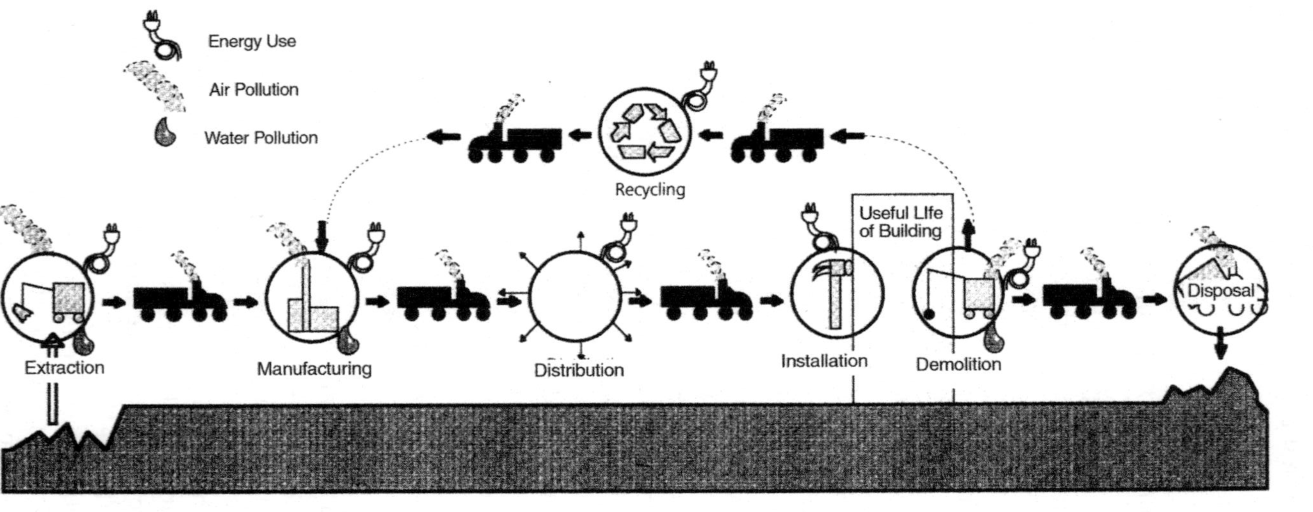

Figure 14.4 Product life-cycle assessment. New York City Department of Design and Consruction, "High Performance Building Guidelines."

consequence, power plant emissions relating to building energy use contributes 33% or roughly 2.5 billion tons of all carbon dioxide emissions, the heat-trapping gas implicated in global warming.[23] It is through this very local, everyday activity of powering our facilities that we unwittingly participate in climate change that is global in scope and irreversible over long time horizons.

Even though national policy makers have yet to commit to specific carbon dioxide reduction measures on the grounds of the scientific uncertainty of real climate change, and the presumably burdensome costs of curtailing emissions, the financial sector — the banking and insurance industries who will be the front line business casualties of climate change — is beginning to emerge as a key player in promoting energy efficiency, renewable energy technology, and sustainable growth policies. These industries are taking the green building challenge seriously. It is recognized that currently available technologies (better insulated buildings, higher performance windows and efficient mechanical systems) could cut commercial building energy use by 60% *and do so at a large net savings rather than increased cost.*[24] The building industry as well — our nation's third largest, with an annual outlay of $800 billion in investments — should recognize the critical role it can play in reducing this nation's carbon dioxide emissions (as part of national reduction necessary under the Kyoto Protocol) through means which are likely to be nearly free or relatively low-cost.

We are just now beginning to understand the high costs of below par performance in yet another critical realm, namely our building's interior environments. The U.S. Environmental Protection Agency has ranked poor indoor air quality as among the top five environmental risks to public health. It claims that unhealthy air (which may be 2 to 8 times *more contaminated* than fresh air) may be found in up to 30% of all new and renovated buildings.[25] Nationally, we may pay a price ranging from $10 to 60 billion in combined health premiums, absenteeism and annual productivity losses due to sick building syndrome and building related illnesses.[26]

Building performance today still falls short of where it could be with some cost-effective course corrections. Many industries now recognize that sound economic and environmental choices, made together, are not only compatible actions, they also produce *value-added end results.* As the real estate community begins to share this perception, high performance building practices will become market driven. "Environmentally-sound" will be perceived as a source of economic advantage.

May the force be with you! — energy efficiency examined

Today's world-view of energy efficiency is quite distinct from the energy conservation mentality of the 1970s which we associate with lower levels of comfort in our homes and places of work. The energy efficiency model of today involves benefits and not sacrifices. For high performance buildings, energy efficient design begins with the methodical reduction *by design* of the building's heating and cooling loads — those imposed by climate and those generated by people and equipment. With all loads minimized, mechanical systems are then selected with the highest output for the least fuel consumption. The new efficiency means obtaining the optimal performance of each contributing component of the building both individually and in interaction with other energy-consuming systems — air conditioning, lighting, domestic hot water, etc. This is the practice of "design integration." Through integration of systems, energy-efficient practices can provide excellent return on investment and support other successful outcomes such as building health and occupant well-being. Today, practices also embrace the use of renewable energy technologies which reduce our dependence on fossil fuels.

Siting and passive solar design

Building energy use is influenced by many siting and architectural decisions. Projects must begin with studies of solar orientation, site features, and shading elements such as other buildings and exterior landscaping. These variables can have significant impact on the heating and cooling loads imposed on the mechanical systems through the treatment of the building skin. Beneficial solar gains and daylight in winter can be admitted through windows, clerestory or skylights, appropriately sized and located. The capacity of a wall or floor material to store the sun's heat energy, known as its thermal mass, can be manipulated to passively heat the building by using this solar gain. Conversely, passive cooling techniques employ appropriate solar shading on openings and the use of natural ventilation (such as the stack effect of hot air rising to an exhaust high in the building) or perform night-time air exchanges (night purge), thereby providing natural cooling or savings obtained by shifting cooling loads to off-peak hours. Simple cost-neutral techniques such as use of light-colored materials for roofs and large adjacent paved areas can deflect incoming heat energy and reduce cooling requirements. So can well placed deciduous vegetation. Used together, these passive solar design strategies can go a long way to providing virtually free comfort benefits.

The building envelope

Some of the most direct opportunities for savings on building energy use will be obtained by improving the configuration and assembly of materials that enclose the building and that provide thermal and moisture protection — its roof, walls, foundation slab. Choice of materials and construction detailing can greatly increase thermal efficiency and cut energy waste for the life span of the building. Maximizing the resistance to heat flow or "R" value of roof and wall assemblies requires attention to the type and placement of insulation. Sheathing and vapor barriers also play a critical part in eliminating condensation and "thermal bridging," which is the creation of unwanted paths for heat energy to enter or exit the building. High insulation values in wall sections can easily be undermined by faulty details which create thermal bridges or air infiltration at structural connections and at window and door openings. The building industry has made significant gains in introducing high performance window assemblies, or super windows — triple-glazed insulating units (high R-value) sometimes gas-filled, usually with spectrally selective glazing in tints which admit daylight and coatings which reduce heat energy transmittance.

Daylight and daylight controls

Today, building designers must virtually rediscover the advantages of daylighting, and regain a familiarity with manipulating daylight that we lost soon after the turn of the century. Conventional electrical lighting design in commercial buildings today consumes between a third and one half of a building's total electrical operating cost. Good daylighting integration may offset that amount by greater than 50%. Some key prescriptions for illuminating a building through daylighting are as follows: place window openings high in the room to throw natural illumination deeper into the space. Use "light shelves" — reflective horizontal surfaces placed at a window opening in the upper portion of a window — to bounce incoming light energy up onto the ceiling and distribute illumination deeper into the room. Ceilings with high reflectance surfaces can then redistribute this light as indirect lighting. Designers also need to integrate electric lighting through the use of continuously dimming ballasts (or stepped dimming) which adjusts lighting levels to offset the daily variation in natural illuminance.

Lighting and equipment

High performance lighting strategies tailor lighting levels and quality to suit occupancy patterns. They use direct or indirect lighting with supplementary task lighting to boost moderate ambient light levels as required. Lighting techniques should consider high–output fixtures (high-efficiency lamps and electronic ballasts) designed for maximum visual comfort in protection against glare. The building should definitely incorporate lighting controls such as occupancy sensors, time clocks, and daylighting dimming photocell controllers (Figure 14.5).

With daylight then minimizing electrical lighting loads, internal heat gains will be substantially reduced, and smaller-sized cooling equipment may result in first costs savings. For further efficiencies, other major office equipment, such as computers, copiers, etc., (known as "plug load") should be selected with attention to the heat they produce.

Load management, right sizing, and controls

After controlling building heating and cooling loads through intelligent lighting and envelope design, the next critical issue is sizing central equipment correctly. This is based on three principles — knowing what the

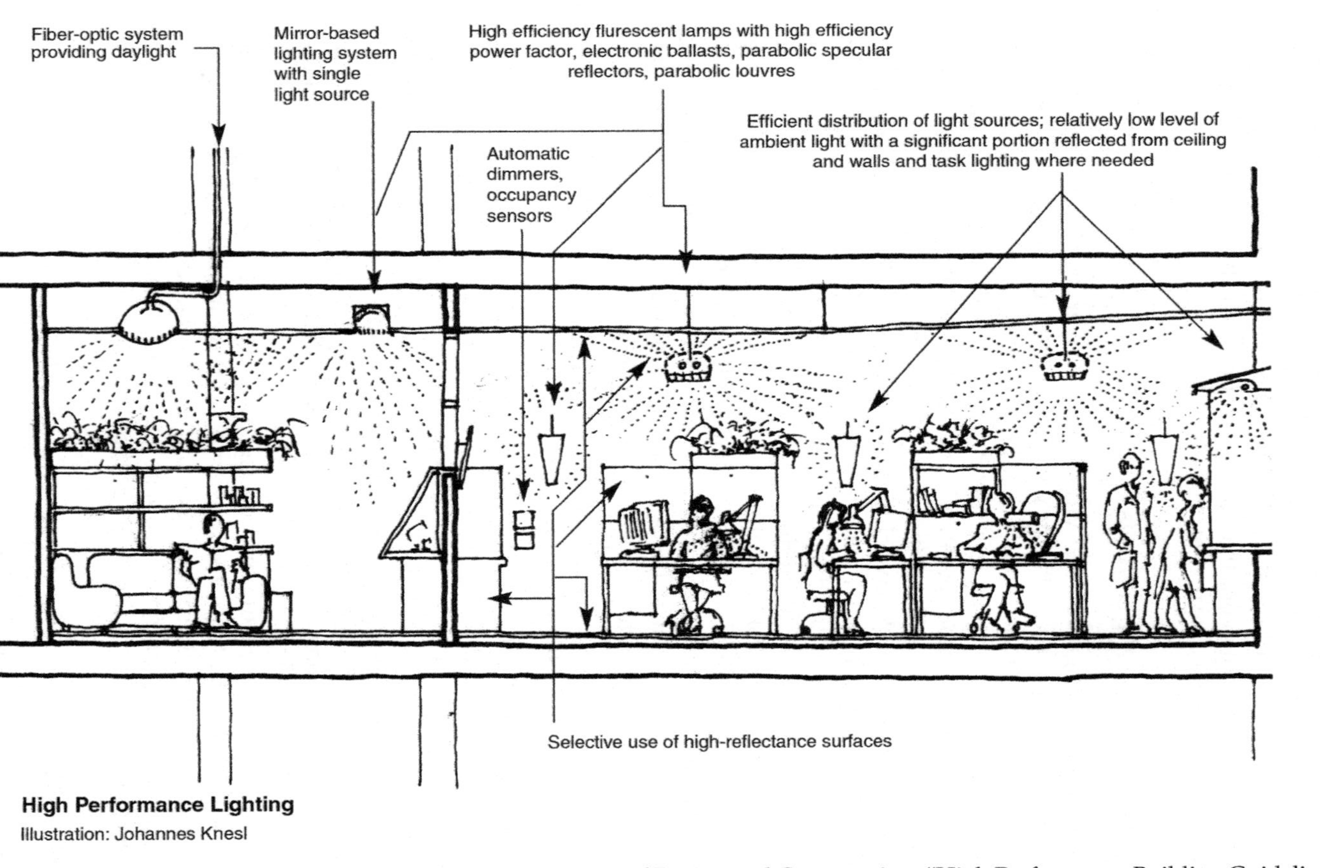

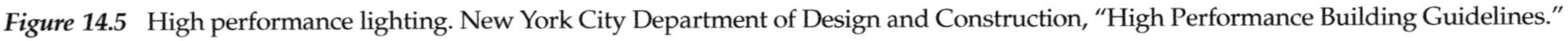

Figure 14.5 High performance lighting. New York City Department of Design and Construction, "High Performance Building Guidelines."

internal heat gains will be from lighting and equipment, understanding how these loads will be diversified or varied based on how the building is actually used, and finally sizing equipment to operate efficiently. Designing for "diversity of occupancy" means understanding that not all spaces need to be designed as though fully occupied all of the time. A school for example, is unlikely to have each space — cafeteria, library, auditorium, gym and all classrooms — full all day long, since students move through each in the course of their activities. Many designers inappropriately base their sizing calculations assuming peak load conditions will exist in all spaces simultaneously.

Most commercial and institutional buildings gain significant internal heat from office equipment, heat discharged from lighting, as well as heat given off by the occupants themselves. Without loss in visual quality, high efficiency lighting has significantly reduced its contribution to the building's internal heat gain. Designers today may safely use a design allowance of less than one watt per square foot instead of the two or three watts assumed for lighting under most energy codes. Accurate estimating of these equipment heat loads — which comprise on average between 15 to 20% of the total cooling load for commercial office buildings — will allow for correct sizing of cooling systems and significant capital savings in terms of chiller size, together with equivalent downsizing of ducts, fans and motors related to cooling equipment.[27]

In the end, it is critical that engineers design and select equipment for typical operating conditions, namely when equipment is operating *at part load,* which is a large percentage of the time. Oversized chiller equipment for example, can cause occupant discomfort through poor humidity control and large temperature variations. Peak load conditions occur relatively infrequently, only when temperatures are at seasonal extremes and all spaces are fully occupied and all office equipment is in use. Central mechanical equipment should be selected that remains efficient over a wide range of load conditions. Modular elements should be considered, such as modular boilers or a series of smaller identical chillers, or even units of varying size. Alternatively, systems using variable speed drive during part load conditions will allow a single chiller to operate more efficiently at part load.[28]

New methods of air delivery are being introduced which impact how we think about heating or cooling spaces. Underfloor air delivery systems (raised flooring with flexible ductwork) can provide for more individual control over workspace. These systems may also simplify future spatial reconfiguration. In large public spaces such as atria or high-ceilinged corridors, the idea of delivering heated and cooled air low in the space *directly to where people are* avoids the inefficient expense of tempering the full air volume of those large spaces. This is one of the concepts behind "displacement ventilation" strategies. Computer modeling tools such as "computational fluid dynamics" can assist engineers to understand air convection patterns against outside walls, and to determine the best way to induce air movement through a low velocity delivery system usually low in the space.

Managing both comfort and operating cost in a building through the incorporation of direct digital control (DDC) systems can support ease of operation, ultimate efficiencies in central boiler and chillers, as well as remote distribution systems like fans and pumps. Controls which can monitor the contribution of lighting and office equipment to the overall building loads further help to decrease energy use through this systems integration approach.

"Design integration"

It is the practice of integrated engineering that takes energy efficient practices and technologies into the realm of high performance. An integrated approach to design focuses all the design disciplines — electrical, mechanical, architectural and plumbing — simultaneously in interactive decision-making on energy use in the building. Integrated design can be supported by computer energy modeling programs, such as the Department of Energy's DOE 2.1E. Energy modeling simulates the design's response to climate and season at a particular geographic locale. It quickly explores cost-effective design options for the building envelope and mechanical systems by predicting energy use of these various alternatives in combination.

The road to renewables

In the next millennium, solar and wind power, fuel cells and biomass — varieties of what are called renewable or alternative energy sources — will likely give traditional fossil fuel energy systems a run for their money. While this technology is only just approaching market-readiness in this country, it has become institutionalized in Germany and many other western European countries where it is perceived not just as a commercially alternative energy source, but as a qualitatively superior one. Designers of high performance buildings should consider building-integrated renewables alongside (but never in lieu of) the more traditional terms of energy efficiency described above. In this way, building owners can participate in readying American markets for renewable technology, while improving their own long-term operating balance sheets as well.

Renewable energy technologies enable us to harness incoming solar energy or tap geothermal or earth-stored energy — both vast and inexhaustible supply sources — in a non-polluting, non-resource depleting manner. It is estimated that in 20 days, the earth receives as much incoming solar energy as is stored in its entire reserves of natural gas, oil and coal![29] One of the most versatile of these emerging technologies is the photovoltaic (PV) cell which uses semi-conductor technology to convert sunlight into electrical energy. PV cells, which are usually comprised of glass substrates with wafers or thin films of silicon, are wired together into arrays to produce direct current. Advances in technology have dramatically decreased the first costs and made the PV cells compatible substitutes for typical building materials — roof shingles, spandrel panels, solar shades — which may further offset costs. In Japan, companies are today producing homes with silicon roof tiles that support most of a family's electrical needs. Wind energy has few urban applications, but installations of modern wind turbines are cost effectively replacing some fossil fuel utility generation in California. Denmark now produces 7% of its electricity from wind power.[30]

Geothermal technology is based on the principle that while air temperatures in many climates may vary seasonally, the temperature of the earth remains relatively constant. Emerging heat pump technologies enable the transfer of heat energy from the ground to a building, or in the cooling mode, taking it out of the building and sinking it into the ground. As an alternative energy source, biomass may be harvested to produce liquid or gas fuels such as ethanol or methanol in many parts of the country where there are large quantities of agricultural wastes (corn or grain crop surpluses) or from energy crops such as willows and poplar. In cities, waste-to-energy applications might someday utilize vast quantities of organic pulp and paper waste, if local environmental concerns for air pollution could be solved satisfactorily.

A technology not strictly a renewable one but deemed most promising today as a transitional, clean energy source, is the fuel cell. The fuel cell combines hydrogen and oxygen without combustion to generate electricity, producing hot water as a useable by-product. These fuel cells are small, silent, and efficient devices which are becoming commercially available. Buildings which operate 24 hours and have large hot water needs, such as hospitals, dormitories, etc., will find them increasingly cost competitive.

Clean energy offered by renewable technology, which could soon be brought about by a combination of private investment and government incentives (tax credits, trading schemes), can enable us to eventually retire fossil fuel technology and its carbon contribution to climate change. It offers us unprecedented opportunities for enterprise and invention. But it will still no doubt require a surge of political will to reform today's energy culture and transform the energy markets. In the meantime, owners and developers today should first ratchet up their building's performance first through the use of integrated, energy efficient technologies, and then consider the opportunities for running the building on integrated renewable energy devices.

Good housekeeping

The animation of the building begins with ribbon-cutting, as customers, tenants, operators, and the general public are ushered across the threshold. Their footfalls track soils and particles into the property; their persons and goods introduce varieties of contaminants indoors. As human activity starts to regulate the building's metabolism, a housekeeping regimen becomes compulsory. A building designed with preventative housekeeping strategies will enjoy a legacy of *built-in economies* in its operations and the management of pollution. Measured on an annual square foot basis, average cleaning cost can quite closely approximate total energy costs. In a high performance project, these activities, like energy, could be expended much more efficiently. Equally important, reduction of bio-contaminants and noxious gases may altogether avoid sick building syndrome, improve indoor air quality, and support overall occupant well-being.

Brighten your "black holes"

Housekeeping productivity gains can be obtained through preventative design strategies that cut down on intense housekeeping activities like floor and surface cleaning labor. This involves keeping most of the dirt and soils from entering the building in the first place. 85% of particulate matter (or most avoidable material) walks itself inside hitched to human conveyors. This represents an opportunity for a large avoided cost as, on average, about 30 to 50% of a building's cleaning budget is dedicated simply to mopping and vacuuming.[31] A proper entry sequence will eliminate unwanted material: rough-textured exterior pathways kept free of plant or other debris will scrape larger matter from foot traffic while metal grating (dimensioned deep enough for two full strides) and additional interior walk-off mats, can capture fine soil while drying shoes. In addition to reducing cleaning frequencies, frictional surface wear from grit is also decreased, increasing the longevity of material finishes.

The same logic applies to housekeeping of general finish materials — reducing cleaning time and frequency through careful selection of patterns and textures, and avoiding too diverse a palette of materials with varying cleaning protocols. In a high performance building, there will necessarily be some trade-offs against other desirable green features. For example, while daylighting strategies are prejudiced towards light-colored finishes, housekeeping ones lean towards darker finishes. The same trade-off goes for dust havens such as light shelves and shading devices, not to mention architectural molding and trim, which will compete for dust-mop attention.

Housekeeping also involves the simple matter of evacuating trash from a facility on a daily basis. Almost 30% of a typical cleaning budget goes to the removal of wastes.[32] Designs which facilitate routine performance of some portion of this function by occupants and/or custodians will save on custodial service costs.

Bathrooms, kitchens, and other high traffic areas often undergo multiple daily purges using caustic cleaning agents. Uninterrupted, non-porous, smooth-surfaced, patterned, round-corned surface materials are desirable for floors and counter surfaces for ease of cleaning and lasting aesthetics. Faucets and fittings should be positioned to prevent splashing and fixtures and stalls should be wall-mounted. The resource efficient principle for good housekeeping is to simplify the cleaning activities — make activities repetitive, involve continuous motions with the fewest possible chemical products and equipment. And consider the janitor's perspective! He (she) will have wanted you to properly size and locate drains, electrical and vacuum outlets, and make sure to vent janitorial storage closets. These few means can express themselves in total building wellness.[33]

Gimme shelter? — besting infestations of pests!

Classes of animals who literally find their niche among us are the creatively opportunistic creatures known as commensal or "synanthropic" pests — those vertebrates (bats, rats, mice, squirrels) and arthropod invertebrates (insects, roaches, ants) who historically have colonized our buildings. Not only unaesthetic and unsanitary, pest co-habitation invites damage to material goods and even structures. They are also major public health concerns as transmitters of microorganisms, allergens, biologic pollutants, and carriers of viruses.[34]

Preemptive pest management is an instrumental feature of healthy high performance buildings. Pest access to building usually occurs through fresh-air intakes, apertures in the envelope (windows, doors) and unsealed wall systems — voids in wall systems from poor detailing and/or construction. Sound strategies bar animal access and harborage by design: sealing with appropriate materials, for example, cement mortar, sheet metal, steel wool, caulking and foam sealants. Smaller insects may be controlled by caulking joints and crevices using caulking compounds (silicone or latex) as well as conventional weatherstripping particularly around wet areas and areas where food is handled. Gaps in ducts, holes in walls, floors, pipe trenches and chases must be considered as access and transit points.

Green 'n clean

Finally, we probably underestimate the importance of selecting the right cleaning products. Most commercial agents have traditionally contained solvents, formaldehydes, and may release CFC's. New products are coming on line which are environmentally preferable (biodegradable) water-based alternatives. Proper dilution can not only create economies, but these "greener" products are ideal from a chemical sensitivity perspective, and make cleaning safest for the janitors and more pleasant for the occupants.

References

1. Roodman, D.M. and N.A. Lenssen, Building Revolution: How Ecology and Health Concerns are Transforming Construction, *Worldwatch Paper 124*, Peterson, J.A., Ed., March, 1995, 52.
2. Romm, J.J. and W.D. Browning, *Greening the Building and the Bottom Line: Increasing Productivity Through Energy-Efficient Design*, 1994, 8.
3. Watson, R.K., *Case Study of Energy Efficient Building Retrofit*, 40 West 40th Street, Headquarters of the National Resources Defense Council, Inc., March, 1990.

4. Ibid, 196.
5. Froeschle, L.M., The Healthful and Efficient Renovation of Ridgehaven Green Office Building, in *Environmental and Economic Balance: The 21st Century Outlook,* Technical Papers, Washington, D.C., 1997, 231.
6. Nicklas, M.H. and G.B. Bailey, *Energy Performance of Daylit Schools in North Carolina*, Raleigh, NC.
7. Ibid, 199.
8. Hathaway, W.E., J.H. Hargreaves, G.W. Thompson, and D. Novitsky, *A study into the effects of light on children of elementary school age: a case of daylight robbery,* Alberta Education, Policy and Planning Branch.
9. Romm, J.J. and W.D. Browning, *Greening the Building and the Bottom Line: Increasing Productivity Through Energy-Efficient Design*, 1994.
10. Figure of 30 percent is from a 1984 World Health Organization committee report, cited in EPA, OAR, "*Indoor Air Facts No. 4: Sick Building Syndrome,*" Washington, D.C.
11. Levin, H., Best sustainable indoor air quality practices in commercial buildings, *Environmental Building News*, May, 1998.
12. Roodman, D.M. and N.A. Lenssen, Building revolution: how ecology and health concerns are transforming construction, *Worldwatch Paper 124,* J.A. Peterson, Ed., March, 1995.
13. National Audubon Society, Croxton Collaborative, Architects. *Audubon House: Building the Environmentally Responsible, Energy-Efficient Office,* New York, N.Y., 1994, 134.
14. Froeschle, L.M., The healthful and efficient renovation of Ridgehaven green office building, in *Environmental and Economic Balance: The 21st Century Outlook,* Technical Papers, Washington D.C., 234, 1997.
15. Fishbein, B.K., *Building for the Future: Strategies to Reduce Construction and Demolition Waste in Municipal Projects*, INFORM, Inc., June, 1998.
16. Rocky Mountain Institute, *Green Development: Integrating Ecology and Real Estate,* John Wiley & Sons, New York, NY, 1998.
17. Crosbie, M.J., Green architecture epitomized: Bethel's Duracell building makes excellent use of surroundings, *Hartford Courant*, March, 1966.
18. Hellmuth, Obatat Kassabaum, Inc., *Sustainable Design 2*, Winter, 1999.
19. Romm, J.J. and W.D. Browning, *Greening the Building and the Bottom Line: Increasing Productivity Through Energy-Efficient Design*, 1994.
20. Gelbspan, R., *The Heat is On: the high stakes battle over Earth's threatened climate,* New York, 1997, 98.
21. Roodman, D.M. and N.A. Lenssen, Building Revolution: How Ecology and Health Concerns are Transforming Construction, *Worldwatch Paper 124,* J.A. Peterson, Ed., March, 1995, 22.
22. Ibid., 210.
23. Ibid., 210.
24. Ibid., 213.
25. Lippiatt, B. and G. Norris, "Selecting Environmentally and Economically Balanced building Materials: *National Institute of Standards and Technology Special Publication 888*, Gaithersburg, MD, 1995, 37.

26. Ibid., 215.
27. Komor, P., Space Cooling Demands From Office Plug Loads, *ASHRAE J.*, 41, December, 1997.
28. Sullivan, C.C., Integrated Design, Key to Efficiency, *Energy User News*, 23:8, August, 1988.
29. Brower, M., *Cool Energy: Renewable Solutions to Environmental Problems*, Cambridge, MA, 1992, 40.
30. Flavin, C., *Clean air in a breeze: the sun and wind are generating solutions*, Energy Futures, Natural Resources Defense Council and Uncommon Sense, Inc., New York, 1998.
31. Ashkin, S.P., *Green and Clean: The Designer's Impact on Housekeeping and Maintenance*, Environmental and Economic Balance: The 21st Century Outlook, Proc. U.S. Green Building Council/Am. Inst. of Architects Conference, 1997, 186.
32. Ibid., 216.
33. Ibid., 216.
34. Frantz, S.C., Architecture and commensal vertebrate pest management, in *Architectural Design and Indoor Microbial Pollution*, R.B. Kundsin, (Ed.), New York, Chapter 11.

Additional Reading

American Institute of Architects, *Environmental Resources Guide*. John Wiley & Sons, 1997.

Environmental Building News, Brattleboro, VT: E Build, Inc.

Green Seal, Inc. *Greening Your Property*, Washington, D.C.: Education Institute (EI), American Hotel & Motel Association, 1996.

Hellmuth, Obata + Kassabaum, Inc. *HOK Database for Healthy and Sustainable Building Materials*, Washington, DC: Hellmuth, Obata + Kassabaum, Inc. 1997.

Kalin and Associates, Inc. *Greenspec*. Newton Centre, MA: Kalin and Associates, Inc., 1996.

New York City Department of Design and Construction, *High Performance Building Guidelines*, 1999.

Leclair, K. and D. Rousseau, *Environmental by Design: A Sourcebook of Environmentally Aware Material Choices*. Vancouver, BC: Hartley & Marks, Ltd., 1992.

Lyle, J.T. *Regenerative Design for Sustainable Development*. New York, John Wiley & Sons, 1994.

Public Technology and US Green Building Council. *Sustainable Building Technical Manual: Green Building Design, Construction, and Operations*. Washington, D.C., Public Technology, Inc., 1996.

Rocky Mountain Institute. Green Development: Intergrating Ecology and Real Estate. New York, John Wiley & Sons, Inc., 1997.

Small, S., (Ed.), *Sustainable Design & Construction Database*: Denver, CO: National Park Service, 1995.

Triangle J. Council of Governments, Design Harmony Architects and Abacot Architecture. WasteSpec: Model Specifications for Construction Waste Reduction, Reuse and Recycling. Research Triangle Park, NC: Triangle J., Council of Governments, 1995.

Watson, D., (Ed.), *The Energy Design Handbook.* Washington, D.C., The AIA Press, 1993.

chapter fifteen

Construction labor in the urban setting

Francis X. McArdle

A view from above a high-rise residential construction site in a city such as New York, London, or Hong Kong immediately calls to mind the purposeful behavior of an ant colony. A hundred or more workers are scurrying from task to task, in what appears to be chaotic patterns over the course of an eight-hour day. But every two days a new floor will appear and the structure will move that much closer to the sky. Questions arise: where do the workers come from; how are they organized to execute the work? What are the challenges facing those who need construction workers?

The organization, development, and use of labor on an urban construction site is quite different from that in almost any other industry. For most urban workers, whether they are in a large American city or a third-world urban environment, work each day is a repetition of what was done the day before, in organizations created to market replicated work for profit. The assembly line encountered each morning in an automobile factory will be changed very little from that left the night before. The supermarket checkout counter will be the same each day. The worker producing a daily or weekly television series will perform the same tasks each day. For the first nine decades of this century, most workers kept their employment relationships very stable. Even as work and residence locations separated during the Twentieth Century through the development of and impacts of zoning ordinances and mass transportation, workers stayed with the same firm for their entire careers. Only layoff-producing economic upheavals, such as a major economic downturn or the collapse of an industry, would produce changes in the employment relationships for most workers.

Construction employment is not like that. With a very few exceptions, construction projects are unique and relatively short-term experiences. Each site has individual characteristics of soil and location. Each design is unique

0-8493-7486-3/01/$0.00+$.50

when married to its site, even if it looks like a similar project undertaken elsewhere. There is simply no analogy in construction for a career in a new automobile plant, for example. A very high percentage of those who started the production systems at the Tennessee Saturn plant or the Alabama Mercedes plant will finish their careers at that plant, with some lasting thirty years. It is unlikely that many of the construction supervisors and workers who built those plants were on the site for more than twenty percent of the short time that it took to build those plants.

Construction projects produce a product with a work force that knows that it must move on when the project is ready for use. The transitory nature of construction work on any given project is one of the organizing themes of the labor issues in construction. The project developer is free to perform the next project with the same set of contractors or to find an entirely new set. The contractors are free to examine the value of the business and the desirability of staying in the business. And the workers are free to move on as well, to new employers or new geographic areas where there is work. The continuing relationship's cost is always evaluated against the economic value received. The impermanence of construction employment affects every aspect of the business, from the time horizons of the construction employer to the objectives of the construction worker.

Development itself tends to be distributed unevenly from a time perspective, responding to the moods of the economy rather than to the desire of construction workers for steady employment. Downturns in the economic cycle, idling production capacity and making developed space available more cheaply than the cost of new construction, leave the construction industry on a downturn, with many workers leaving the industry to find the jobs that will support their families. The boom-and-bust nature of the urban construction industry development cycle affects the longevity of average industry employment and leaves the industry always having to develop a substantially new workforce when the bust side of the cycle turns positive once again. Construction work, therefore, tends to pay more for labor to attract new workers during the boom cycle.

In the United States, and perhaps everywhere and for all times, short-duration construction contracting has always been a high-risk/high-reward enterprise, particularly open to entrepreneurship. The cost of market entry is very low, particularly in urban areas, where smaller and more precise subcontracting has become a constant goal of the project developer. Supplies and equipment are readily available for limited credit. Emerging construction employers often have already developed technical skills by working in construction as an employee. What will distinguish them from other workers is a drive to create financial opportunities for their families.

Project developers, whether in government or the private sector, have always had a great interest in obtaining the most work for the lowest price in order to encourage new firms. The smaller the subcontract work portion, the smaller the firm that can take the work and perform it successfully. The smaller the firm that can undertake the work, the more likely there will be firms competing for the work. The more competition, the lower the profit margin accepted by the winner and the smaller the total cost of all subcontracts. The more competitors and the smaller their size, the greater is the risk that the successful proposer will underprice the work to be done. The developer can then benefit from the uncompensated investment of new capital by the firm that underprices the work. Of course, the burden of coordination and control for the developer increases as more firms are hired. Generally speaking, the tendency in construction in urban environments over the last century has been to reduce the size of the subcontracting package and hope for the best on coordination, control, and the replacement of small subcontractors who run out of working capital. Construction entrepreneurship (until the emergence of the Internet) was seen as one of the easiest and most open ways to build family wealth for those without access to capital or the professions.

In the United States, as in many other urbanized nations, the differences found in the organization and employment of construction labor are explicitly recognized in the labor laws that have been adopted over the last seventy years. Unlike the broad industry-wide reach of most unions, construction trades are very narrow in their grasp, each covering only a few of the tasks performed on a construction site. So the construction industry, particularly in the urban setting, is characterized by multi-employer bargaining. Construction trade unions are allowed to enter into pre-hire agreements with employers to regulate the supply and conditions for labor, unlike their industrial counterparts. The understanding of these laws and the way they are changing is key to how much labor issues will affect the construction industry in the next millenium.

In most urbanized nations, there are national union organizations structured to represent the interests of workers in the construction industry. The reach of these national union organizations varies with the strength of national policy on workforce representation and the economic influence of the unions on worker training. In many European countries prior to the EU, the national unions were very strong because they monopolized worker training and supported worker representation. In the United States, where the commitment to representation is weaker and training control is limited, the national union reach is limited.

The national unions are not all the same, even as they stand together as construction industry unions. The differences are important, and become clearest at the smallest unit of union organization, the local. The two strands in construction employment in the urban setting become obvious, because of their disparate origins and their impacts on the construction labor force. One strand is that of the skilled trades, the specialists who install the electrical systems, the plumbing systems, the mechanical systems, and steel structures of buildings. These groups of specialists, each organized and educated differently, are really urban construction guilds, with connections in ethos and skills to the guilds of constructors in medieval Europe. The other strand is that of the laboring trades, involving the employment of people of undifferentiated skills and abilities, who assist the specialists or do the work that sets the stage for the specialists.

The skilled trades strand has roots that go back most clearly to the cathedral building and castle building of the Middle Ages in Europe. The contemporary connection to medieval building is clear because it involves the organization of the education and employment of free men.

The urban construction guild today represents a lineal connection to the development and emergence of artisan guilds in the construction of the cathedrals and castles of the Middle Ages. All forms of medieval work were organized around the guild. The guild controlled the marketing of products, the production of products, and the education of those who were to become producers. The guild maintained the quality standards of production through the education process and the quality control over that which was produced. In the development of these roles, the guild was furthering the economic development of the urban setting in which it existed. Travelers or traders would rely on the stature of a particular guild or guilds in making their decision on the acquisition of goods. An area with high guild standards could find that their production was traded far beyond their immediate catchment area; declining guild standards could lead to the demise of an urban center.

The positive values of the guild did not come unfettered. Entrance into the guilds was carefully controlled, and often proved to be very exclusionary. Cities often licensed their guilds and gave the guild legally binding jurisdiction and enforcement powers over work. Often only those already inside the guild could successfully propose relatives or siblings for guild apprenticeships. Training was often restricted to those favored by the guild leadership, resulting in the perpetuation of leadership. And guilds often restricted the use of raw materials to those sanctioned by the guild. This kind of "closed" guild was antithetical to the constant immigration that urban areas have experienced since the dawn of history. In medieval Europe, these restrictive guilds encountered surplus farm labor and restless entrepreneurs in constant battles. Those battles continued until the emergence of the industrial city, in which large-scale production facilities overwhelmed most of the guilds and made them anachronisms.

But in urban construction today, the themes of conflict and control of labor in medieval Europe can be found alive and well. The medieval construction guild assured the builder of a house or the builder of a cathedral that the work would be of a certain standard. Membership in a construction guild, such as the stonemasons, would allow one to travel to work opportunities, with the guild membership certifying in advance of employment that production standards would be met. But guild membership was restrictive, and often the growing demands for construction employment could not be met. Often, budding but alien entrepreneurs would be excluded from membership and forced to find work elsewhere. Those excluded from the guilds often formed a parallel construction industry, working at the fringes until they could force their way into the main stream. From among these excluded often came the workers and organizers of the "laboring trades."

Can one find a parallel today to the medieval guild? Consider the role that is played today in New York City by Local 3 of the International Brotherhood of Electrical Workers (IBEW). In New York City, the city licenses many trades, very much like medieval cities. One cannot legally act as a master electrician or a master plumber or a master rigger in New York City unless one obtains a city license to do so, irrespective of prior experience and expertise. The testing for these licenses is done infrequently to test standards that favor those who have trained in the locals most closely associated with the trade. The test standards are very conservative, often emphasizing specific skills that have long fallen out of favor in the practice of the trade; only the insiders can possibly know the arcana necessary to pass the test. Local 3, the local branch of the IBEW, controls the intake of electricians to its apprenticeship program, matching those that it will train to its perceptions of the market demand for labor. Through that control, Local 3 is able to affect the number of master electricians who will be allowed to practice as employers in the New York City marketplace. Through its collective bargaining agreements with those employers, Local 3 further extends its control over the marketplace. Electrical fixtures installed in New York City by Local 3 must be approved and certified by Local 3. Everyone involved in an electrical construction project, from the master electrician through the journeymen to the truck drivers and secretaries are all members of Local 3. Local 3 is a medieval guild in contemporary garb, setting high quality standards for production of a potentially dangerous product and sharing the same tendencies to control and exclusion that precipitated struggles in the medieval city.

The broad reach of Local 3 over the electrical construction tasks encountered on an urban construction project is replicated, to some degree, by other local units of the IBEW in other cities or the locals in New York City that cover other skilled trades. The hoisting trade (or operating engineers) tries to control all of the hoisting done on a project. They will make every effort to insure that only members of the local will work on the hoisting on the project. Members of the local have an inherent advantage in seeking licensing to hoist from New York City. What began as training and quality control has changed.

They work hard to preserve the traditional jurisdiction of their local, even as technological changes no longer require that compressors be tended at all times to assure continual functioning and that hoisting machinery be constantly lubricated.

Their jurisdiction, however, does not cover all hoisting on the site. The operating engineers encounter hoisting done by the stone setters, those who put on the exterior stone facades of buildings. There, the hoisting of materials may be done by the stone setters themselves or by a unique trade, the stone derrick men. The territorial edges of jurisdiction can become more uncertain. The introduction of new materials or new methods precipitates conflicts between the locals over jurisdiction, the question of which trade union (or urban construction guild) will perform what work.

In many urban areas over the last century, the membership in the urban construction guilds has been a central issue in local labor politics and local elective politics. Why? The financial rewards of construction employment are substantial. Urban construction represents an opportunity for anyone who is physically able to earn a living. As well, the ability to control guild membership involves the ability to provide employment for one's children, particularly male children, and to leave them a legacy that has long-term value: membership in the union guild. This membership has often been passed from father to son for generations. The members of the construction guilds believe in that legacy and resist the intrusion of outsiders in the running of the guild and the selection of its members. The ethos and organizing values of the guilds have always come up against the push of the outsiders, whether in 1916 in New York City or in 1998 in Miami. The faces of those pushing out the invaders and those invading have been different over time. But when those who are pushing in get in, they adopt the ethos of the guild as strongly as those that went before them. The thrust of the push is the same.

Urban America has experienced continual waves of immigration from other countries during the Twentieth Century. Also, there has been a constant migration from rural areas within the United States to the cities as a result of rural surplus population and of the urban pull. These waves of in-migration to cities have been under some degree of control, but in a country with 5,000 miles of porous borders, control has never been absolute. Whatever the source of the immigration, urban construction projects have been magnets for those migrants, and the construction industry has looked to these people to replenish the industry workforce during upturns in the development cycle. Construction has usually paid more than the industrial work requiring equivalent skills; the work tends to be dirty and outdoors, rather than indoors and clean. Urban construction work is temporary. Everyone wants the security of permanent work and will not stay in construction unless they can get into the guild. And that's where the problem originates.

The attitudes of contractors and developers play a role here as well. Immigrants are perceived to work harder and better than those who are born into the urban culture where the work is being performed. I believe contractors across the globe share this attitude. Contractors in American cities believe that emigrants from India work harder than American-born labor. Contractors in New Delhi believe that workers from rural areas in the Punjab or Uttar Pradesh work harder. A conscious effort to support immigration thus develops. Some of the strongest opponents to the United States anti-immigration statutes of the '20s came from the heavy construction industry in New York City. The contractors knew then that if the supply of immigrant labor was cut off, recruitment to expand the workforce would be harder and that the costs of maintaining production on the construction site would rise.

The impact of immigration on the fortunes of the construction industry was measured very clearly in New York City in 1916. The city decided in the years before the outbreak of World War I to undertake many major investments in its urban infrastructure, including substantial expansions of its subway/elevated transportation system and its water supply system. (These expansions precipitated an investment in the development of new office building space as well.) As usual, the construction industry looked to the immigrant workforce to staff the expansion. When projects were bid out for construction in 1915, unskilled labor cost $2 a day, and projects were bid accordingly. But World War I intervened and substantially altered the workforce equation. New emigration from Europe stopped, as workers in Europe were conscripted into the national armies. And, the call-up of reserves precipitated out-migration, as over 300,000 individuals went home to answer the call to war. Suddenly, with construction contracts in hand and bid, contractors had to find new labor supplies. Within two years, the cost of labor rose to $3.75 per day, and, while promises were made, no relief was provided to the contractors. By the time that the dust had settled in 1919, almost all of the contractors who had done work on the subway and water supply expansions were bankrupt due to the rise in wage rates.

The most dramatic confrontations between the urban skilled trade unions and the city immigrants in the United States have come over the last four decades, as blacks who migrated from the rural south, pushed out by the mechanization of southern agriculture, have sought jobs in urban construction to improve their family economics. The exclusion of blacks from guild membership might have been simply one more step in the traditional exclusion and incorporation model, but three elements changed that. The first was the commitment in 1964 by the wider society to the goals of equal citizenship and access for black America. That commitment, accelerated by the death of President John Kennedy, produced both laws and an enforcement mechanism, the Equal Employment Opportunities Commission (EEOC), that would be used over time to crack open the control of the guild system in urban construction. The second element, perhaps developing from the same impulse that precipitated the national commitment to civil rights, was the increased role of the federal government of the financing of urban construction projects and its requirement for fair employment. The third element — only now being fully appreciated — is the new emphasis on technical and college-based education as the key to upward familial mobility.

Blacks, like most previous outsiders to construction, have never been absent from the urban construction scene. But until quite recently, they were found most frequently among the "laboring trades," the groups that installed concrete (but didn't finish it) or dug ditches. Look at any photograph of workers constructing New York City's water tunnels and one will find black faces. One will find similar faces among those that dug ditches or poured concrete. But they weren't to be found in any numbers among the highly trained guild trades of urban construction. This exclusion was both based on deeper prejudice and in the routine exclusion by the guilds of any outsiders of any race or nationality. As has been mentioned, union membership had traditionally been passed on to only the family and friends of those who were already members. The passage of the 1964 Civil Rights Act and the subsequent expansion of the powers of the EEOC provided a tool to combat the exclusionary practices of the construction guilds, precipitating a clash that still exists between the idea of openness and access for all Americans and the idea of the guild membership as an asset, like an acre of land, that can be passed on to one's heirs. In those clashes, one found "liberal America" arrayed against "the hardhats."

The closed nature of the construction guild trades, whether in Philadelphia, Chicago, or Boston, was well known to all. The internal operating ethos of the guild was not as well understood. The direct assault on the urban construction guild system, which began in the early 1970s and paralleled the assault on urban school discrimination, was never matched with an assault on the suburban housing segregation system, so guild members reacted very strongly to what they perceived to be a discriminatory assault against them. Most guild unions consistently resisted the inclusion of blacks under court order. In New York City, an EEOC action against Local 28 of the Sheet Metal Workers Union commenced in 1972 but is still ongoing, with findings as recently as four years ago that the local was still discriminating against its minority membership. The guilds closed in on themselves. If they had to take in minorities, they would only take a few, until the next enforcement action. If they had to be trained, then they would be trained for the lowest skilled jobs. If they had to be sent out to jobs, they would be sent last and they would be laid off first.

The increasing federal financing of urban construction projects gave the national commitment to equal employment another venue for attack. Every federal construction project was utilized, and is utilized, to advance the object of equal employment, with specific construction employment goals set for minority participation. In New York City, perhaps because of its role as a media center, those goals were set by specific trade, unlike anywhere else in the country, so that the total hours of participation required could not be achieved without the disruption of the guild system. The unintended effect of this approach was to isolate minority participation to the public sector jobs with federal participation, often called "checker boarding," where minority employees would be redirected to job sites one step ahead of the inspectors. Contractors could thus be in compliance on various jobs citywide with only a few minority employees.

Organizations of black workers pushed strongly for inclusion in the urban construction workforce. Their objectives, like all of those who went before them from other ethnic groups, were access to the jobs that could pay a decent living to those without the best of formal training. As other minority groups emerged in the urban area, whether Hispanics from Puerto Rico or from the Hispanic southwest, they too formed groups to push for workforce inclusion. Job site violence became common, as "coalitions" of minority workers demanded work. But workforce inclusion didn't necessarily end the clashes. Entrepreneuring minority organizers saw an opportunity to transmogrify inclusion demands into money-making opportunities. Group leaders would approach construction companies, demanding the inclusion of workers from their group and the exclusion of unaffiliated minority workers, threatening violence if the demand was not met. Accidents resulted in the employment of a "coordinator" from the group, who would, in emulation of the padrone model, supply labor and fight off any who challenged. The "coordinator" would receive both a salary from the contractor and a "commission" from every worker placed. Contractors who were confronted with the choice between the modest cost of the coordinator and the high cost in lost production and money from violence, given very slow police response, hired the coordinator so long as he or she was able to control the situation. When there was a successful challenge to the power of the "coordinator," the winner became the new coordinator. The "coordinator" system has been attacked effectively by the New York City police as a racketeering enterprise, with the jailing of four major "coalition" organizers. The emergence of new leadership has been tempered by the activities of the police and the growing demand for labor, muting the cry of the coalition leader.

The assault on the guild system by the EEOC and the employment requirements on federally aided projects have had a substantial impact on the opening up of guild membership. But, equally significant, has been the shift in the availability and access of college and technical education in the United States over the past 35 years. When the Civil Rights Act was passed in 1964, college education was still largely unavailable to the children of construction workers in the United States. Fifteen years later, with the massive expansion of the public university system and the development of the community college model from its roots in California, college education is now available to and chosen by the children of the working classes of America. And the work is there for those who choose to follow that route. The effect on the guild system has been dramatic. Fathers who could not foresee anything but a trade legacy for their children in 1964, now have the opportunity, and their children the desire, to see their children go to college. So children choose to go to college, and they leave the trade behind. This has fundamentally altered the guild system in most urban environments in North America. The resentment of intruders is still very strong and the commitment to the guild heritage is still very strong, but not every child is destined to follow his father's footsteps. And, particularly at this point in time, the guilds are recognizing that their futures as organizations will depend on their intake of members who look nothing like themselves.

chapter sixteen

Women in construction

Lenore Janis and Evelyn Mertens

Contents

The fastest growing women-owned business of the 1990s 233
Emerging from the recession: the growth of public/private partnerships 234
Will women crack the "concrete" ceiling? 234
Real estate — development: a last frontier for women? 235
The post-war era: the invisible woman 236
Engineering and architecture: a lonely path in the early years 236
The 1970s — women's liberation and affirmative action 238
Business booms and women thrive in the 1980s 239

"Women suppliers were new at first. Now there are women at every facet of the job site. You won't find that quizzical look of 'what is she doing here?' anymore when you see a woman engineer," said Henry Estrada, vice president of community relations with Tishman Construction Company. He says that the climate has changed.

Improvements in educational opportunities also deserve credit for the increased numbers of women in the construction industry. As Cooper Union's Engineering Dean, Eleanor Baum points out, "Society became more technologically oriented during a time when the women's movement made it okay to have two-career families." Women began signing up in the 1970s and more so in the '80s at technical schools and departments within other colleges and universities for construction management courses.

Martin Sandler of NYU's construction management diploma program notes that women have been well represented among the student body over the past 15 years. The percent of women students has been as high as 17 to 20%, "in times of economic growth," and as low as 10 to 12% during periods

0-8493-7486-3/01/$0.00+$.50

of economic uncertainty. The confluence of positive forces led to a development that would have appeared bizarre in the 1960s: the appearance of women project managers.

The project manager, (PM), is the heart and soul of the construction industry. The PM is the conductor of the orchestra, coordinating every aspect of the project from site to budget. The PM is the lifeline for management. Initially the thinking was that women were not trained or experienced in this capacity and could not stand the rigors of working in the field. There was some truth in this, but it was a Catch-22 situation: how to get trained if no one is willing to train you, and how to get hired if no one will hire you?

Anne Avenius, currently a highly placed project manager for Barney Skanska, managed to defeat the odds. "Construction interested me right away," says Avenius. "Instead of seeing a pile of papers move from here to there, after a year or two you see an actual building you've constructed."

Though Avenius never questioned her career options in the early 1970s — nurse, mother, or teacher, — when she graduated in 1973 with a bachelor's degree in teaching, she found the market glutted. While working as an office manager for an architectural firm in Manhattan, she began to take courses in construction management at NYU, New York Polytechnic Institute, and Pratt Institute.

She worked her way up until she became a project manager. "There were cat calls at first when I visited a job site," she recalls. "But I convinced them I knew what I was doing and went on with my business." She has worked at several major construction companies in New York City and today is known as a heavy hitter in the industry.

Debra Shore currently project manages the Fort Lauderdale and Hollywood International Airport Terminals 2 and 3 refurbishment for Parsons Brinckerhoff. She came into the business when conditions were somewhat more favorably inclined to women. She says, "In the mid-'80s, if you were bidding for a multibillion dollar government job, it was a plus to have a woman in any key position — a competent woman who could do the job, that is."

Shore stresses the competency requirement noting that there's no room for style over substance in construction. "It's a cutthroat business where a major deal can be lost on pennies. No one is going to take any chances on you if you can't perform."

While working as PM at Terminal 1 at JFK, Shore was on the job ten hours a day, working with superintendents, foremen, and subcontractors. She felt she was treated with respect because "clients and my company had faith in my ability."

What about when the desk is turned and Shore needs to seek WBEs as subcontractors? "Now it's no longer a problem to find qualified businesses. We find them by seeking good companies. The largest contract we have is with a woman-owned business whom we would have chosen regardless of the goals."

The fastest-growing women-owned business of the 1990s

Today, the U.S. Census Bureau and the National Foundation for Women Business Owners (NFWBO) rates construction as the fastest-growing women-owned business: from 1992 to 1999, a growth of 68%. The number of women-owned firms has risen from 119,687 in 1987 to 391,900 in 1999. Still, these figures need to be viewed in context. In 1994, *Dun & Bradstreet Report* said that there had been a 19.2% increase from 1991 to 1994 in the number of women-owned construction firms nationwide. Yet the percent of women-owned firms was only 4% of the total number of firms.

Lina Gottesman grew up in a family-owned stone setting business. She named her company, formed in the late 1980s, Altus Metal and Marble Maintenance, after a mythological Greek bird, an altus, symbol of excellence. She says, "There's a certain thrust put forth for WBEs that you have to be better than your competitors."

She believes that being a WBE helped her get her foot in the door. "But after that, you have to develop a reputation as a quality company."

Gottesman suggests that the reasons for the growth spurt for women in the industry are varied: "There's money to be made. Also, it's a challenge, and women want challenges. Women want to compete with the boys. We're cut out from that competition at a young age — when we're told that the Little League is not for girls."

Figure 16.1 PWC President Lenore Janis at a project on the 59th Street Bridge, Manhattan.

Emerging from the recession: the growth of public/private partnerships

Fortunately, many WBEs survived the recession of the late 1980s and early 1990s. When the construction industry took a nosedive, fewer people were hired, more were fired, and businesses failed or weren't formed. Women were, once again, particularly hard hit due to the age-old "last hired, first fired" dictum. Since they were the new kids on the block, they were the first to go.

And yet, partly because more sizable numbers of women had already been incorporated into the workforce as managers, subcontractors, and professionals, the recession hurt, but didn't eliminate women from the picture. In addition, even by the late 1980s, support networks, both formal and informal, were firmly in place and playing an active role. Estrada notes that "PWC acts as a sort of banner under which all businesses meet. Major companies have the opportunity to come into contact with more WBEs than ever before." The Society of Women Engineers has also been a great help to women.

Affirmative Action helped WBEs survive the recession of the early 1990s because much of the work available was publicly funded. As the industry began to pick up its pieces in the mid '90s, it became clear that there had been a tremendous amount of overbuilding in New York City in the previous decade. Much of the new work therefore became not new construction but interior renovations and repairs, and redevelopment of old factories and warehouses for commercial and residential use. When building resumed in the mid to late 1990s, much was through public/private partnerships, a creation born during the last decade of the millennium. Those joint ventures and consortiums were formed to build billions of dollars of new airport and light rail facilities to meet the needs of the twenty first century.

Unfortunately, Affirmative Action suffered a major blow through a largely unpublicized reversal in June of 1998 when New York City Mayor Rudolph Giuliani allowed the City's Affirmative Action laws to expire quietly under the "sunset" clause. Therefore, there are currently no mandates for work done through wholly city–held funds. Says Gottesman, "It's a sad setback for women. Its effects are starting to be felt already and will become only greater with time."

Will women crack the "concrete" ceiling?

Of course, the other side of the coin is that more and more women are forming WBE firms because they've hit the so-called "concrete ceiling" in trying to move up the ladder in major companies. The possibility of becoming CEO of one of the titans of the industry is still slim.

Susan Hayes is president and CEO of Cauldwell Wingate Company, Inc., a construction management firm with annual revenues of $50 million, making it a mid-sized firm. Hayes notes that "there are very, very few women on the executive level on the building side of construction. Many more are in architecture and design. In fact, some of the most powerful women in the industry are on the architectural side."

Hayes explains that "women came late to the industry." Until recently, one achieved executive status by beginning in the trades and moving up the ladder. "Now the industry has changed dramatically. The level of sophistication is greater and true management skills are required. The top person must be able to lead the organization. The client wants someone with business sense as well as building knowledge." Now she encourages other women, including architects, to enter the construction management side of the profession.

Real estate — development: a last frontier for women?

Is there a woman in construction who has never said to herself "Will we ever see a female Donald Trump?" The corporate climb is only one horizon for women in the industry to ascend. Real estate development is another.

Lois Weiss, real estate journalist with *The New York Post* and the president of BetweenTheBricks.com, a real estate information Web site, says, "There's a future for women developers because there isn't a past. They've been on the sidelines or accessories to men. But so many have served as right hands in the last 10 or 20 years, it's looking more and more likely that they will want to do it on their own." Several women were instrumental in the redevelopment of the Times Square area. As with project managers, "for a long time women lacked experience in this area. Now, they're getting the experience."

The major obstacle is, once again, funding. It is extremely difficult for any new developer, male or female, to obtain financing for major projects entirely through private investors. Therefore, women are finding potential in development projects either subsidized by the government or financed, at least in part, through major fundraising campaigns. Some of the areas in which women developers are beginning to show success are museums, municipal redevelopment projects, and affordable housing.

Construction is, after all, about as macho as the workplace gets. It is a locker room elevated onto steel girders, a world of hard hats and steel-toed boots where women were long forbidden to tread. Ladies did not build buildings or bridges, for that matter. In truth, the facts prove otherwise. There are reports that women engineered the sewers of Pompeii in ancient Rome. And, when John Roebling became incapacitated during the erection of the Brooklyn Bridge in the 1870s, his wife, Emily Warren Roebling, stepped in as chief surrogate engineer and facilitated the construction through to its completion. That piece of anecdotal history is now emerging from the realm of trivia, as more people recognize Mrs. Roebling as a true pioneer and role model for future generations.

Yet it took nearly 100 years for women to begin to gain a presence in the industry. Although Betty Friedan's landmark work, *The Feminine Mystique,* broke ground in 1963 and Affirmative Action laws began to appear on the books during the L.B. Johnson Administration, there were no visible changes in gender on construction jobsites for at least another 15 years.

The post-war era: the invisible woman

Of course, women were working, even playing active roles, in construction, in the 1960s and even before, particularly in "mom and pop" businesses. But they played their parts behind closed doors: not seen, seldom heard. For example, a lucrative family business is bequeathed equally to a pair of siblings, male and female. The sister is not content to sell her share and beats a hasty retreat to hearth and home. She stays in the office and makes sure the numbers line up while brother ventures out to do the glad-handing and back slapping. Her role is no less vital to the continuation and growth of the company.

The late Sarah Potok, one of the founders of PWC, served as comptroller of Lasker Goldman, a family run business, a major construction company in New York City, from the 1950s through the 80s. A strong-willed, spirited woman, she started to work at age 16 in a jobsite trailer and developed a reputation as a powerful, no-nonsense player in the industry. Though certainly a trailblazer, she was not entirely an exception to the rules of the game since her work was also conducted primarily *in* the office.

Women in family-owned construction businesses did surface occasionally. Certainly, being part of management helped. One woman remembers that one of her "outside" duties was to deliver the cash payroll to the iron workers on site. She once asked the foreman how the men felt about a woman boss. "As long as they get paid on time, they don't care who's bringing the money," he replied.

Some women entered the field almost by accident. Hired by construction companies in such traditionally female positions as clerks, secretaries, receptionists, and bookkeepers, sometimes, if they worked under a progressive management, they might be promoted through the ranks until they achieved a decision-making role. PWC's executive vice president, Theresa Vigilante, began as a receptionist at Creative Woodworking, once a major architectural woodworking firm on the east coast. Within ten years, she had become corporate treasurer, one of the company's four officers, and the only woman.

Engineering and architecture: a lonely path in the early years

Engineering and architecture are two other avenues in the industry largely blocked to women. Women engineers emigrating from Israel and Eastern Europe were starting to make their appearance in U.S. firms, but only a few native souls dared to brave the tide.

Eleanor Baum, currently dean of engineering at The Cooper Union for Science and Art in New York City, was the only woman in her engineering class at City College in the mid-1960s. "People thought it very strange. At that time, women became schoolteachers or nurses," she says. After graduating, she had no great trouble obtaining a job in the aerospace industry. "It was the Sputnik era, and engineers were needed." Yet once on the job, she felt isolated and missed a support system of women.

The picture has changed substantially in the past 30 years. Baum notes that the greatest numbers of women enroll in environmental and chemical engineering degree programs, areas that offer "a hope of improving the quality of life for present and future generations." She notes, too, that in New York many women see "renewal of the infrastructure as a major problem" and may be drawn to civil engineering.

Engineering graduating classes have grown from 1% women in the 1970s to 15 or 16% now, nationally. According to the Bureau of Labor Statistics, by 1998, 11.1% of all engineers were women, up from 7.3% in 1988. Women chemical engineers numbered 16.5% (up from 12% ten years ago), and 12.1% of all civil engineers and 7% of all mechanical engineers were women, roughly double their percentages in 1988. (Figures on environmental engineers were not available.) Baum notes that three-quarters of all women engineers have a brother or father who is an engineer. At Cooper Union, Baum's vigorous recruitment policies have resulted in an increase from 5% to 38% women in the various schools of engineering since she began her tenure in the early 1970s.

Today, Marilyn Jordan Taylor is one of two women partners (of 27) at Skidmore, Owings & Merrill (SOM), a world-renowned architectural firm. A full 27% (or 65) of the 244 architects at SOM are women.

When Taylor told her family she'd decided to opt for a career in architecture over law, they all, including her father ("my greatest mentor"), encouraged her goals. The only question was: Would she *sound* like an architect too?." Taylor explains: "The image was of a rough and tough world where foul language reigned."

She was one of seven women in her class of fourteen at MIT when she began her training in 1969. But when she completed her studies at the University of California at Berkeley in 1973, 10% of the architecture students were women, a more representative sampling of the national picture. Even today, only slightly over 10% of A.I.A.'s members are women.

Taylor had no problem finding work after graduation. Hired immediately by SOM, (where she had done field work while in school), the stumbling block was client relations. By age 27, she was director of station design for Amtrak's Northeast Corridor, supervising 110 people on a $2.5 billion railroad improvement project.

Taylor recalls, "I was dealing with older men, engineers. I was always the only woman in the room, and you could read the reaction on their faces. Was I the plague sent to destroy them? But I felt a great responsibility to do well. If I made a mistake, I felt that I'd be jeopardizing the credibility of all women. Still, I always thought of myself as an architect first."

It was Taylor's ability coupled with her skills in client relations that allowed her to succeed. "The model partner was a male who not only led the team, but entertained clients. Gradually, though, perceptions changed," she says.

The 1970s – women's liberation and affirmative action

Changes in attitudes had to be, of course, nudged by laws, lawsuits, and lobbying. Women's liberation helped the children of the 1960s choose to forge a career before marrying and raising a family. Yet, while popular image holds the 1970s to be the time when the women's movement was in full swing, the beat often struck a dissonant chord when it came to non-traditional fields.

In 1979, President Jimmy Carter signed an executive order to create a national policy and program for the establishment of Women's Business Enterprises, that came to be known as WBEs. The policy held that "Each department and agency of the Executive Branch shall take appropriate action to facilitate, preserve, and strengthen women's business enterprises, and to ensure full participation by women in the free enterprise system." It was an important victory, and yet its application was still limited. The language was sufficiently general and devoid of specific goals or mandates and allowed for much circumvention.

Some general contracting companies set up "sham" WBEs that weren't really run by women. That hurt the efforts of legitimate WBEs. In addition, government agencies in charge of enforcement were understaffed and could not monitor good faith efforts of general contractors to seek WBE participation.

Dickie Sykes, corporate director of minority business development, and equal employment opportunity officer with Morse Diesel International, Inc., notes that "legislation won't change attitudes. People hired people who looked like them, in gender and ethnicity. It took years for Affirmative Action to evolve into something of significance."

And, as Sykes points out, "The fear was that the women and minorities would be unqualified. And, of course, those who were would hurt the others. But the talent and know-how of the minority business enterprises (MBEs) and WBEs who survived, finally won out. Affirmative Action doesn't mean unqualified; it gives qualified people opportunities. Mandates get them through the door, later you can hear: 'wow, they're good.'"

It was late in the 1980s when a small but growing number of women, finding themselves tantalized by career potentials in construction, yet thwarted in their efforts to move ahead, founded Professional Women in Construction. The founders were a mixed group composed primarily of: entrepreneurs who had grown up surrounded by family owned businesses; and women working for midsized construction companies who felt trapped in traditional jobs that left them no room to advance. Their purpose was to create public awareness of the existence of women in the industry, a support system among themselves, and a centralized voice to advocate in their own behalf.

PWC's first event was a two-day convention held in New York City in November 1980. Topics included "Women and the Law," "Financing a Small Business," and "What the Women's Business Enterprise Program Means for You." The consensus was, the construction industry stonewalls women.

One of the first aims of PWC was to advocate for specific goals for WBEs. In the early years, minority organizations outnumbered women's groups and were more successful in their lobbying efforts. This resulted in goals set at 5% to 15% for MBEs.

Reflecting a similar logic, the U.S. Small Business Administration in 1979 and 1980 issued loans that were capped at $150,000 for minority business enterprises (MBEs), and at $25,000 for women. The reasoning behind those low loans for women, called mini-loans, was, "How much does a bagel-baking business or a beauty parlor need to get off the ground?"

Perhaps the greatest obstacle to the growth of women subcontractors in construction came from the general contractors. At first, the Associated General Contractors (AGC) reacted with extreme hostility to any effort by the government to decree whom they should hire. But by the mid-80s, the landscape had so changed that every major company had an EEO office. Further, more and more women-owned companies were establishing themselves as highly qualified and competitive.

Yet being hired was less than half the battle. It was extremely difficult for a woman to maintain a new business or even an inherited company, since it was virtually impossible for her to obtain loans or get credit from suppliers. Sadly, this led to the demise of many a family-owned business when "pop" died. A widow would usually have no recourse but to sell, often below market value after her husband's death... unless, of course, there were sons to take over.

The particular needs of so-called "rich widows" were finally recognized by the federal government in the 1980s, due, in part to lobbying efforts by PWC. The government argued that a woman inheriting a thriving business shouldn't need the protection of Affirmative Action laws. But this wasn't a question of haves and have-nots. PWC and others maintained that women would not have been able to continue to run a business without the enforcement of Affirmative Action mandates.

In 1983, PWC also succeeded in its efforts, along with other women's groups, to establish stated goals for the inclusion of WBEs on publicly financed projects. The newly elected Governor of New York State, Mario Cuomo, set goals for WBEs at 5%, and his administration actively became involved in enforcing the laws. Hiring women, at last, had become a "must" rather than a "maybe."

Business booms and women thrive in the 1980s

By the 1980s, all forces converged to help women gain at least a toehold in the industry. The mandates forced companies to seek a diverse work force.

That helped both minorities and women gain access. In addition, business was booming and work was flowing. When the construction industry thrives, everyone in business benefits from the bounty. There was, in fact, a lack of personnel, and qualified women were sought after by major companies.

Mike Mennella, executive vice president of Tishman Construction, New York, says that "Affirmative Action inspired many women to venture into their own businesses. It gave them an added incentive to compete and the time to catch up with already established businesses. We concentrate on hiring talented and experienced subcontractors, and today, many WBEs can bid with the best of them so we don't worry about meeting goals. On the management end, we've been scooping up trained and knowledgeable women because of their background."

Construction requires guts and brains, but not brawn. This was a revelation 30 years ago. Today it's almost an accepted truth. And so, will women shatter the taboos of generations, and build buildings? Says journalist Lois Weiss, of *Real Estate Weekly*, "As big as they can get."

chapter seventeen

University of Washington case study

Jon Marmor

After decades of strained relations, the University of Washington (UW) and neighboring communities are beginning to get along, just as three major projects are about to hit the "U" District.

That sound you hear rumbling through the UW and the Seattle neighborhoods that surround the campus? Could it be the collective grinding of teeth by the 50,000 people who live within a mile of the UW, and the 50,000 or so UW employees, faculty and students who populate the campus. The reason? The region is going to be hit with three of the messiest, biggest-impact projects and all at about the same time. Brace yourself:

- Light rail is coming.
- The Seahawks will play at Husky Stadium in 2000 and 2001.
- Construction of the William H. Gates School of Law and possibly up to 3 million square feet in new buildings on campus.

If you make your home near the UW or work on campus, life could be a total headache for years. Consider: six years of construction (beginning in late 2000) on a subway line on the campus side of 15th Avenue N.E.; 71,000 Seahawk fans jamming into Husky Stadium on Sundays in the next two autumns; and a cornucopia of possible construction projects -- ranging from the new William H. Gates School of Law to possibly putting a lid over Pacific Avenue N.E. — to accommodate 10,000 more people on campus in the next decade.

"It's pretty overwhelming," says Louise Little, president of the Greater University District Chamber of Commerce and personnel director at the University Book Store. Yet, despite the coming disruption, there is a strange feeling permeating these parts.

0-8493-7486-3/01/$0.00+$.50

For the first time in recent memory, the UW and the surrounding neighborhoods have been acting, well, neighborly toward each other. Years of rancor, mistrust, suspicion, and contempt seem to be subsiding. And just in time, too.

Oh, things are not perfect. There are still lots of ongoing squabbles. "But the situation is much better than it used to be," says Steve Sheppard, the supervisor of the Neighborhood Programs Division for the city of Seattle, who for years has been involved in dealings between the university and surrounding resident groups.

Perhaps the best example of how relations between the university and community have changed was the stunning resolution of a potentially nasty situation — the siting of a new UW indoor athletic practice facility. It looked like yet another big university vs. community battle.

The $26 million facility will be about 80 feet high and 150 yards long, holding about 95,000 square feet. The original site was north of the Intramural Activities building, parallel to Montlake Boulevard. But fearing it would create a "canyon effect" down the busy thoroughfare (because of the hill on the other side of the street), residents said it was unacceptable.

Just as tempers were beginning to flare, Barbara Hedges, the UW athletic director, turned out to attend several community meetings to see how the situation could be worked out. Her mere presence at those meetings blew many people away.

"Community concerns are of great importance to us all," she explains. "It became very apparent to me that the concerns of the community regarding the site were much greater than anyone had anticipated. I listened very carefully to what they were saying to us."

So she went back to the drawing board and came up with a new site on Walla Walla Road on the Lake Washington waterfront behind Husky Stadium. That alternative site will cost the athletic department approximately $2 million more and delay the project by a year because of the need to address new environmental concerns. But the UW gets the facility it wants, and preserves harmony with the community.

That move generated such goodwill that community organizers have begun writing governmental officials in support of the UW's application for permits in that sensitive wetland area.

The solution of this matter has been hailed in all quarters. "This solution is very symbolic," says Sheppard. "People listened. That made for a tremendous change in University-community relations. It built trust and understanding. The people involved learned they could talk, listen and work things out, even if the answers weren't quite what they wanted to hear."

But most of the talking between the two sides for many years was not exactly friendly. Bad feelings started back in the 1950s and 1960s, when communities felt the University and city were in cahoots to make sure the UW could gobble up all the land it wanted along Portage Bay so it could expand. Those moves generated lawsuits and a distrust of the UW that lingers to this day. "There's always been the feeling that the UW would buy up more land to preserve the sylvan feel of campus," says Fred Hart, proprietor of La Tienda, a shop on University Way N.E. "The university is here to provide education. It should not be looked at as a national park."

Those hard feelings run the gamut, from legal action and threats to stop the expansion of the UW Medical Center and keep the UW from developing the Southwest campus to ongoing clamor to tear down the four-foot-high concrete wall bordering the campus side of 15th Avenue N.E. Just about everyone agrees that wall is an affront to those living and toiling outside the lush campus environs. "The wall is psychological as well," says one local resident. "It's part of the unfriendliness of the campus, a little thing that shows the University's attitude and insensitivity to those outside." Getting along has always been a prickly proposition here, given the number of players involved. There's the university. The city. Community councils from Montlake, Ravenna Bryant, University District, University Park, Wallingford and Laurelhurst. There's the Greater U District Chamber of Commerce. Various other quasi-governmental councils, and, last but not least, the 50,000 residents in and around the campus. All have their own interests at stake.

The first attempts at ironing out the cantankerous relationship between the University and surrounding communities began to take shape in 1976. That's when the city of Seattle and the UW forged a Memorandum of Understanding to create a committee that would serve as a "viable mechanism for information dissemination and citizen input" and help put a stop to increasing enmity between the University and surrounding communities.

As a result, the City University Community Advisory Committee, known as CUCAC, was born. A coalition of 16 members from seven community councils, UW administrators, faculty, students and a city appointee, it was set up to advise the city and university on the orderly physical development of the university, adjacent community and business areas, and protect those areas from adverse effects of university and city actions.

CUCAC has opened a line of communication between the parties, and is playing a vital role in the UW's current Campus Master Plan, a document that proposes adding 3 million square feet in new buildings and physical refinements in and around campus for the years 2002-2012.

The proposed development responds to the forecasted growth in student enrollment, and is comparable to the physical development that has occurred on the Seattle campus during the past 10 years.

New buildings aren't the only potential result of the Campus Master Plan. Planners are looking for opportunities to strengthen connections between central campus and other areas both on campus and in the community. One such idea under consideration is putting a "lid" on Pacific Avenue N.E. to treat the busy street as a "seam," rather than an edge between the sections of campus. Some residents remain skeptical of the UW's intentions. "I have lived here all my life, I went to the UW, but the U is not a good neighbor," said one campus neighbor who has battled with the UW for years. There are huge problems with the university building baseball fields, soccer fields and this proposed indoor practice facility on landfill in

Portage Bay. That used to be a garbage dump. An earthquake could be a real problem. The Urban Horticulture Center is built on a peat bog dump. What if an earthquake hits? And the university is first and foremost a university. Why would it even consider having the Seahawks play in Husky Stadium? That is going to make our neighborhood a nightmare. The U is supposed to be dignified and not have an emphasis on making money off pro sports." But the UW's role as a neighbor has been appreciated more and more. Local activists cheered the relocation of KUOW's offices into the old JC Penney Building on University Way N.E. because of the stability and prestige it brings. When 11th Avenue N.E. was torn up last summer for construction, the UW allowed Metro to park its buses in campus parking lots. The UW School of Architecture will be involved in a facelift and redesign of the beloved University Heights Center for the Community's grounds. And when Chris Curtis, a 1973 university graduate, started up the University District Farmers Market back in 1993, she benefited from the help of students in the public relations and marketing program at the UW School of Communications. Her farmers market is now the state's largest.

"While some people have felt the university couldn't be trusted," says Patty Whisler, who heads up The University Way Planning Group, an organization devoted to improving the area adjacent to campus, "there's a lot more understanding of the university as a benefit to the community, and to our lives."

"In the past," adds Bridget A. Chandler, the UW's assistant vice president for regional affairs, "neighborhoods were afraid of us as the 800-pound gorilla that was going to expand beyond our boundaries. But now we talk about being partners, of understanding each other, and creating situations where we both win.

"That is what being neighbors is all about."

Appendix

The international perspective

Patricia J. Lancaster

There is not a place in the world today where a construction project could be built that is safe from the possibility of its being stopped by special interest groups. Information, technology and policy abound, as do special causes and the funding for them. In what are now called developing countries the country and/or its leaders may be non-consequential, but they are building, and nothing can be built without funding, and the funding source is often as vulnerable to pressure as the country itself is vulnerable to pressure from the funding source. This is not to say that non-controversial projects with independent funding will not go ahead, just that now as never before there is always a chance that your project may become controversial without any desire or action on your part. Your deal may appear seamless until it starts to unravel. Even here in our own country, Washington State (by introducing new legislation) negated a negotiated and signed design/build agreement for a new highway. In developing countries, the problem is compounded by the fact that you may be in contract with an entity that does not exist next month, no matter how stable they seemed the day they signed. Since there is so much flux in governments around the world, the future will have funding sources solidly in the middle of our deals. You will not only be building for the King of… or Prime Minister of… but for the President of whatever banking entity is financing the venture.

These financiers are not necessarily out to do the right thing for mankind and the planet; they want to make money. If a project is delayed, money is lost; if there is negative press, revenue is lost. They must operate so that there are no delays, and so that the project does not call attention to itself. It will be up to the design and construction professionals to ascertain the universe of possible concerns and assess the probability of each one of them

0-8493-7486-3/01/$0.00+$.50

causing delays or bad public relations. Construction will only be one part of the services needed to complete a project.

Cross-national concerns and treaties between governments that affect our industry (if you do not count autocracies) are relatively recent. In the 1960s fish were the focus, and actions centered around ocean pollution and migrating fish. But in 1972, the first international environmental conference was held in Stockholm: the United Nations Conference on the Human Environment. This conference either purposely or unwittingly, has laid the groundwork for all subsequent international environment law, and had the support of policy-makers from many nations. Of course, even the agenda for that conference took years to develop and then was amorphous. Generalities were much easier than specifics, and economically no country could afford to enact what its environmentalists purported. While concern was palpable, veracity of all facts was unable to be proven.

The Stockholm Conference passed 109 resolutions in 5 categories:

- human settlements
- natural resources management
- pollutants
- education
- development

The categories each have their major issues, from global warming, destruction of the food chain and depletion of natural resources to providing safe, affordable housing in burgeoning populations, chemical pollutants from manufacturing and training the local workforce to help themselves. Most of these global issues have environmentalists shouting to policy-makers even as they are in disagreement over the facts. Will the greenhouse effect counteract the imminent coming of the next Ice Age? What are the tangible effects of the disappearing rain forests in Brazil? What is the net effect on humankind of a few more extinct species? Scientists disagree among themselves. And after all, bio-genetics may soon be able to re-create these species.

More to the point for our industry, though-incidents and issues are not evaluated in terms of their value or effect on construction, design, real estate and development. Second–growth trees are not as strong as first–growth trees, so many more second–growth trees are needed to build a structure, and the individual members need to be larger — thus affecting aesthetics. Property values along oil tanker routes went down after the Exxon Valdez travesty. Re-building after hurricanes, earthquakes, and bombs pours money and jobs into our industry. In fact, even wars are very good for manufacturing, production facilities, and construction in general.

In the 1970s, the international environmental community convened over and over in various permutations and combinations and passed resolutions that were either powerful and obeyed or not so powerful and given lip service but not obeyed. The players were at war among themselves for credibility and jurisdiction, and their agendas and power rose and fell like a sleeping animal's chest.

In the 1980's, governments were increasingly pressured by their constituents to take stands on environmental issues, and the world's economy became much more of a factor in policy-making. Globally, as well as in the United States, those who are making the policy are neither knowledgeable about nor interested in the welfare of our industry.

The 1990's brought globalization of huge industrial segments with offices overseas for firms that used to be mom and pop shops. And there were many, many purchases of United States-based construction firms by overseas concerns.

The situation today is that there will be an increasing demand for "respect." Acknowledgment and concessions will need to be made to the "top three" issues of pollutants, forestation, and energy efficiency. Since population, economics, and environmental concerns are inter-related — they will all influence the outcome of future construction projects. Since global problems are inextricably linked to individual behavior, individual behavior will be examined ever more closely. Technology will make this examination easier than ever before.

It is likely that agreement will be reached among many nations about what project or impact criteria will trigger the need for some sort of environmental impact assessment, and that policies and programs will result. The effect of these eventualities will be that individual projects will be mandated to come under scrutiny, whereas they may have been able to be built before with no scrutiny. It is also likely that more governments will be pressured to have review(s) be held in the public domain and not limited to internal agencies or those selected by the government.

The biggest issues are: 1) how to affect the outcomes at the local level when the concerns are huge and global, and 2) the on-going competition for jurisdiction at all levels.

Index

A

Above-ground space restrictions, 158
Ad hoc community interest groups, 5
ADA, *See* Americans with Disabilities Act
Affirmative Action, 234, 238, 240
Afro-American Coalition of Construction Workers of City, 63
AGC, *See* Associated General Contractors
AIA, 237
Alabama Mercedes plant, 222
Altus Metal and Marble Maintenance, 233
AMEC Civil Engineering, 7, 10
Americans with Disabilities Act, 3, 120
Amsterdam tree soil, 174
Amtrak's Northeast Corridor, 237
Archaeological investigations, 111, 112–113
Archaeology and construction, 135–136
  archaeological inventory, 139
  background research, 139
  concerns, 137
  consultants, 137–138, 140, 142
  excavation, 141–143
  government concerns, 144
  laws and procedures, 136–137
  library research, 140
  permitting process, 136
  phases, 138
  preliminary field testing, 140
  site evaluation, 141
Architect, 136
Art Commission, 71
As-of-right building permit, 2, 3, 152
Associated General Contractors, 239
ASTM field tests, 85
Atlanta, 77
Audubon Society, 202
Autowalk, 18
Avenius, Anne, 232
Avenue of the Americas, Eleven Hundred, 83, 86–89

B

Back Bay, 28–30
Balfour Beatty Major Projects, 7, 10
Bangkok, 36
Banks Street, 75
Barney's New York, 89–92
Barwick, Kent, 148
Baugher, Sherene, 150
Baum, Eleanor, 231, 236, 237
Belle, Beyer Blinder, 126
Beneficial occupancy, 197–199
Berms, 162–164
  German tree protection standards, 163
  site constraints, 163
  urban limitations, 164
BetweenTheBricks.com, 235
Big Ben, 7, 11, 12, 14, 16
Biochemical oxygen demand, 71
Black & Latin Economic Survival Society, 63
Black & Puerto Rican Coalition of Construction Workers, 63
Black art, *See* Compensation grouting
Board of Estimate Approval, 73, 107
  application approval, 109, 111, 112
  resolution conditions, 109–110
Board of Standards and Appeals, 107, 110, 152
BOD, *See* Biochemical oxygen demand
Boroscope, 120
Boston, 77
Boundary survey, 113
Brooklyn Bridge, 53, 235
Building Department, 130
Building foundations, 80
  permits, 80
  surveys prior to construction, 80–81
Buildings, high performance, 194, 205
  alternative officing, 201
  build less options, 201
  cross-disciplinary collaboration, 205
  daylight controls, 211

energy efficiency, 209–210
envelope design, 210, 211
geothermal technology, 215
green developers, 205
guidelines, 200, 204, 208, 213
heating and cooling loads, controlling, 210, 211, 212, 213, 214
housekeeping, 215–216
lighting and equipment, 211
pest infestation, 217
telecommuting, 201
thermal bridging, 210
universal sizing, 201
Bureau of Labor Statistics, 237

C

CADD drawings, 120
California Bearing Ratio, 174
Camp Letterman, 143
Campus Master Plan, 243
Canal Street, 70
Canopy structural stability, 158
Carey, Hugh, 150
Carter, Jimmy, 238
Cauldwell Wingate Company, 234
Center City, 84
Central Artery, 78, 82
Central Artery/Tunnel Project, 37
Certificates of Inspection, 114
Chandler, Bridget A., 244
Channel Tunnel, 11
Charing Cross Station, 11, 14, 18
Charles River, 29
Charles River Dam, 47
Chesapeake Bay, 149
Chicago, 79
Chicago River, 124
Cistercian, 13
City archaeologist, 150
City Archaeology Program, 144, 152
City College, 236
City Hall Park, 142
City Planning Commission, 71, 72
City University Community Advisory Committee, 243
Civil engineering, 27
Civil Rights Act of 1964, 228, 229
Civil War, 143
Clean Air Act, 2, 3. *See also* Environmental regulation
Clean Water Act, 2. *See also* Environmental regulation
Columbia University Medical School, 95, 97, 99
Columbia-Presbyterian Medical Center, 95, 97
Community Liaison Officer, 62
Compensation grouting, 7, 14, 15, 18
Composts, 162
Con Edison, 61
Conditional Negative Declaration, 114
Conrail, 84
Construction in an urban environment, 25–27
geotechnical issues and settings, 25–27, 28
historical perspective, 28–30, 32–33, 35
Construction industry
costs for new construction, 118
cross-national concerns, 246
economic issues, 4–5
environmental measures, 2, 3
federal mandates, 2
hidden costs, 207
international perspective, 245–247
over city streets, 78
political climate, 3–4
preservation projects, 118
product life-cycle analysis, 118
social agendas, 5
Construction labor in urban settings, 221
coordinator, 229
employment opportunities, 226
federal financing, 228
guilds, 224, 226, 228
immigrants, 226–227
influx of blacks from rural south, 227
job site violence, 228–229
laboring trades, 227
minority organizers, 229
national union organizations, 223–224
project developers, 222
short-duration constructing contracting, 222
skilled trades, 223, 224
Construction manager, 136
Construction site facilities, 60–62
Construction under city streets, 79, 105–107
approvals required, 107
building codes, 107, 108, 113–116
environmental quality, 107, 111–113
excavation/foundation permit, 113–114
franchise agreement, 110
land use, 107, 108–110
zoning, 107, 110–111
Contaminated groundwater, 55
Contractor's role, 205
commissioning process, 206

cost considerations, 206–207
site protection plan, 205
specifications, setting explicit, 206
up-front process, 205
Control of Pollution Act, 22
Cooper Union for Science and Art, 236
Cornell University, 175
Corporation Counsel, 71
Craul, Phillip, 168, 170
Creative Woodworking, 236
CRM, *See* Cultural resource management
CSI Division One, 206
CU Soil, 174–175
CUCAC, *See* City University Community Advisory Committee
Cultural resource management, 144, 145
Cuomo, Mario, 98, 99, 239
Curtis, Chris, 244
Cut-and-cover tunnel construction, 35

D

D&CL, *See* District and Circle Line
Daylight, beneficial effect, 197–199
achievement tests, student's, 198
full spectrum fluorescent lighting, 198
low emissivity, 198
reduction in energy costs, 198
DDC, *See* Direct digital control
Dead dog park, 100, 104. *See also* New York State Psychiatric Institute
Delaware River, 75
DEP, *See* Department of Environmental Protection
Department of City Planning, 152
Department of Energy DOE 2, 196, 214
Department of Environmental Protection, 111
Department of Plant and Structures, 71
Depleted groundwater supply, 55
Design process and principles
construction, new, 119
historical building, 118, 119–120, 121, 122, 123–124
integration, 214
rehabilitation project, 119
renewable energy technologies, 214–215
traditional building, 118
Detroit River, 32, 33
Developer, 136
Dinkins, Mayor, 98
Direct digital control, 214
Displacement contour plots, 15, 16
District and Circle Line, 12
Dow, 2
Dunn & Bradstreet Report, 233
Duracell International, 203

E

Earth Pressure Balance, 35
East River, 70
East River Drive, 78
Economic Development Corporation, 99
Economic issues, influence upon construction, 4–5
EDC, *See* Economic Development Corporation
EEOC, *See* Equal Opportunities Commission
EISs, *See* Environmental Impact Statements
Eldon, Ethan, 98
Electrical construction project, 225
Ellerbe Becket, Inc., 94
Emory Roth and Sons, 89
Endangered Species Act, 2. *See also* Environmental regulation
Energy Resource Center, 202
Energy Star Building Program, 196
Environmental hardiness, 158
Environmental Impact Statements, 3, 98, 103
Environmental Impact Study, 20
Environmental Protection Agency, 71, 72, 199
Environmental Quality Review, 114
Environmental regulation, 2
EPA, *See* Environmental Protection Agency
EPB, *See* Earth Pressure Balance
Equal Opportunities Commission, 227, 228, 229
Estrada, Henry, 231, 234
EU, *See* European Union
Euro-American fortification, 142, 143
European Union, 223
European Union Mandate, 2. *See also* Environmental regulation
Excavation challenges, 36
mega-cities, 36
slurry wall method, 40
Extensometers, 15
Exterior wall renovation, challenges, 83
Externalized costs, 194

F

Fair share legislation, 5
Farrell, Denny, 97
Feminine Mystique, The, 235

Fife, Barbara, 103
Fort Washington Park, 95
Fortune 500 companies, 86
Fox and Fowle, 145
Fox, Robert, 145, 150
Freeman, Charles R., 29
Friedan, Betty, 235

G

Gamestetter, D., 173
Geological conditions in Central London, 19
George Washington Bridge, 100, 101
Geotechnical and environmental issues, interrelationship, 46–47
Geotechnology, 27
Gettysburg, 143
Giuliani, Rudolph, 234
Global Conference, 2. *See also* Environmental regulation
Global warming, 209
Gotham Construction, 133
Gottesman, Lina, 233
Government Activities Specialist, 64
Government archaeologist, 144
Gowanus Expressway, 78, 82
Graham Anderson Probst & White, 124
Grand Central Terminal, 82
Grand Trunk Railroad, 30
Greater U District Chamber of Commerce, 243
Greater University District Chamber of Commerce, 241
Green building movement, 203
Green Park, 10
Greening the Bottom Line, 198–199
Greenwich Village, 75
Ground freezing, 52
Ground settlement, 55
Groundwater and construction, 49–50
  control for digging and building, 50–51
  control methods, 51–53
  costs, 54
  dewatering, 55, 56
  flora and wetlands, harming, 56–57
  side effects, undesirable, 54–56
  structure design below the water table, 53

H

Harkness family, 97
Harlem Fight Back, 63
Harlem River, 75
Harris, J. Roger, 158
Harris, Roger J., 158
Hart, Fred, 243
Harvard Square, 56
Harvard Yard, 56
Hayes, Susan, 234
Hedges, Barbara, 242
Henri Bendel Retail Store, 128, 130–133
Henry Hudson Parkway, 73
Henry VIII, 13
HEPA, *See* High efficiency particulate air
High efficiency particulate air, 201
Historical building construction process, 121–122
  addition, 122
  construction procurement, 122
  dismantling and demolition, 122
  intervention types, 121–122
  mason restoration subcontractor, 123
  rehabilitation, 122
  restoration, 122
  site logistics, 123
  subcontractors and craftsman, need, 122, 123
  team, construction, 122–123
Historical preservation, 118
  architect/engineer team, 124
  research, 119–120
  structure, understanding, 119
  survey and assessment, 120
Hog House person, 61
Hoisting trade, 225
Hotspots, *See* Infrared photography
Houses of Parliament, 11
Hudson River, 72, 95, 104, 151
Hudson River Compact, 71
Husky Stadium, 241, 244
Hydrometric analysis, 120

I

IAQ, *See* Indoor air quality
IBEW, *See* International Brotherhood of Electrical Workers
Inclinometers, 15
Indoor air quality
  flush-out period, 201
  management plan, 199
  microbial organic compounds, 199
  off-gassed, 199
  unhealthy air, effects, 199
Infrared photography, 120
Infrastructure
  assessment, 14
  face loss, 14

getting along with existing, 77
green field site, 14
limiting damage to existing, 13
Inspection reports, 114
Internal Revenue Service, 151
International Brotherhood of Electrical Workers, 225
Investment Tax Credit, 86
IRS, *See* Internal Revenue Service
Isaacs, Abraham, 152

J

Jack Train Associates, 127
Janis, Lenore, 233
Jet grouting, 52
JLEB, *See* Jubilee Line Extension Project
Johnson County, North Carolina, 197
Johnson Wax, S.C., 197, 205
Johnson, L.B.J., 236
Journeymen, 225
Jubilee Line Extension Project, 7, 8, 9, 10
environmental protection, 20, 22
infrastructures, monitoring, 18
London locations, 11
noise and vibration, 22

K

Kennedy, John, 227
Koch, Edward, 149
Kohn Pederson Fox Associates PC, 84, 85
Kopinga, 170
KPF, *See* Kohn Pederson Fox Associates PC
Kyoto Protocol, 209

L

La Tienda, 243
Labor Specialist, 63
Lake Washington, 242
Lalique, Rene, 130
Land surveyor, 113
Landmark approvals, 120–121
Landmarks Preservation Commission, 112
Landmarks, approvals for, 120–121
Certificate of Appropriateness, 120–121
local municipal reviews, 121
preservation commissions, 120
Landscape architect, 136
Langthorne Abbey, 13
Lasker Goldman, 236
Leadership in Energy and Environmental Design, 196
LEED, *See* Leadership in Energy and Environmental Design
Lerner, Jill N., 94
Light patterns, 158
Limited, Inc., The, 133
Linares, Guillermo, 97, 99
Lindsay model, *See* Trees and urban construction
Little, Louise, 241
Livingston House trench test, 143
Local 3, 225
Lockheed's Building 157, 196
London, 8, 10
geological concerns in Central London, 19–20
Roman influence, 8, 13
London Bridge, 11, 13
London Traffic Directorate, 10
London Underground Ltd., 10
Los Angeles, 36, 77
Love Canal, 2
LPC, *See* Landmarks Preservation Commission
LUL, *See* London Underground Ltd.
Luna, Maria, 97, 99

M

Maclowe, Harry, 1
Madison Avenue, Six Hundred Sixty Six, 83, 89–92
Madison Square Garden, 8
Manhattan Island, 75
Mariners Museum, 149
Market Street, 84
Master electrician, 225
Material data specification sheets, 64
MBEs, *See* Minority business enterprises
MDWF, *See* Mean dry weather flow
Mean dry weather flow, 74–75
Mellon Bank Building, 84, 85, 86
Memorandum of Understanding, 243
Mennella, Mike, 240
Merchandise Mart Retail Development, 124, 126–127
Messinger, Ruth, 98
Met Life, *See* Metropolitan Insurance Co.
Metropolitan Life Insurance Co., 90
Metropolitan Sewerage Commission, 70
Mexico City, 36
Milstein Hospital, 102
Minority business enterprises, 238
Mixed-face tunneling, 33, 35–36
Mobilization, 59–60
Montlake, 243

Morgan Lewis Bockius, 85
Morse Diesel International, 238
MSDS, *See* Material data specification sheets
Municipal Arts Society, 150
Museum of London, 13

N

National Environmental Policy Act, 2, 3. *See also* Environmental regulation
National Foundation for Women Business Owners, 233
National Historic Preservation Act, 2. *See also* Environmental regulation
National Register of Historic Places, 121
Natural Resources Defense Council, 196
Neighborhood Programs Division, 242
NEPA, *See* National Environmental Policy Act
New York, 36, 69, 77, 79, 82, 136
  labor unions, 225
New York Central Railroad, 71, 73
New York City Landmarks Preservation Commission, 130, 144, 145, 152
sunken ship, 145, 148–151
New York City Planning Commission, 130, 144
New York City Site Selection Board, 71
New York Polytechnic Institute, 232
New York Post, 235
New York State Department of Health, 71
New York State Office of Mental Health, 95, 97, 103
New York State Psychiatric Institute, 93, 97
  environmental issues, 101–102
  expansion plans and strategy, 97–99, 100, 102–104
  founded, 94–95
  Kolb Annex, 95, 102
  neighborhood concerns, 95, 96, 99
  public and private affiliations, 95
  renovation controversy, 95, 97–98
  site and program, 100–102
New York Unearthed: City Archaeology, 153
New York University, 232
NFWBO, *See* National Foundation for Women Business Owners
North American town site, 142
North River Treatment Plant, 71, 72, 74, 75
North River, *See* Hudson River
Norway, 82
NYC Water Tunnel, 61
NYSPI, *See* New York State Psychiatric Institute
NYU, *See* New York University

O

Obata, Hellmuth, 205
Occupational Safety and Health Administration, 63, 64
Odell, Bill, 205
Oldham, John, 95
OMH, *See* New York State Office of Mental Health
Open pumping, 51
OSHA inspector, *See* U.S. Labor Department Compliance Officer
OSHA, *See* Occupational Safety and Health Administration

P

Palace of Westminster, 11
Pardes, Herbert, 97, 99
Parliament Square, 12
Parsons Brinckeroff, 232
Passive solar design, 210
PCB, 64
Penetrometer reading, 174
Penn-Central Railroad Yards, 73
Pennsylvania Station, 8
Pepper Construction Company, 127
Permeation grouting, 52
Pest infestation, 217
  public health concerns, 217
  synanthropic, 217
Phase 1, 139–140. *See also* Archaeology and construction
Phase 2, 141. *See also* Archaeology and construction
Phase 3, 141-144. *See also* Archaeology and construction
Photovoltaic cell, 214–215
PI, *See* Psychiatric Institute
Pipe arch, 20
PM, *See* Project manager
Political climate, influence on construction, 3–4
Portage Bay, 242
Post Office Square, 38
Potok, Sarah, 236
Power at Last, 63
Pran, Peter, 94
Pratt Institute, 232
Presbyterian Hospital, 99, 102, 103
President's Council on the Environment, 3

Product Life Cycle Assessment, 208
Professional Archaeologists of New York City, 152
Professional Women in Construction, 236, 238, 239
Project manager, 232
Psychiatric Institute, 99. *See also* New York State Psychiatric Institute
PV, *See* Photovoltaic cell
PWC, *See* Professional Women in Construction

R

Raised bed planters, 162–164
Ravenna Bryant, 243
Real estate and building industry, worldwide, 195
Real Estate Weekly, 240
Reiss, Warren, 145
Replacement/reconfiguration of multi-lane highways, 78
Resource Conservation and Recovery Act, 2. *See also* Environmental regulation
Ridgehaven Office, 196, 202
River North neighborhood, 126
Riverbank State Park, 74
Riverside Drive, 97, 98, 100, 102, 104
Riverside Park, 72, 97
Rockefeller, Nelson, 72
Roebling, Emily Warren, 235
Roebling, John, 235
Roebling, Washington, 53
Ronson,Howard, 144, 148, 149, 150, 152
Rose Garden, 203
Rubin Organization/Equitable, 85
Russia Wharf Tunnel, 46

S

Safety Supervisor, 63
Sail effect, *See* Trees and urban construction
San Diego, 196
San Francisco, 77
Sandler, Martin, 231
Sarnia Tunnel, 35
SBS, *See* Sick building syndrome
Seaman's Church, 151
Seatle Seahawks, 241, 244
Secant piles, 52
Secretary of the Interiors Standards for Rehabilitation, 121
Section 106 review, 121
SEQRA, *See* State Environmental Quality Review Act
SGI, *See* Spheroidal graphite iron
Sheppard, Steve, 242
Shore, Debra, 232
SHPO, *See* State Historic Preservation Office
Sick building syndrome, 199
Singapore, 79
Six Penn Center, 83, 84–86
Skanska, Barney, 232
Skidmore, Owings & Merrill, 237
Slurry walls, 42, 54
Social legislation, effect on construction, 5
Society of Women Engineers, 234
Sodium Absorption Ratio, 162
Soil bearing capacity, 114
Soil boring logs, 113
Soil-tunnel *vs.* rock-tunnel construction, 32–33, 35
Soils Systems, 144
Soldier piles and lagging, 42
SOM, *See* Skidmore, Owings & Merrill
South Ferry, 151
South River, *See* Delaware River
South Street Seaport, 135
South Street Seaport Museum, 145, 153
Southern California Gas Company, 202
Special interest groups, 5
Spectrographic analysis, 120
Spheroidal graphite iron, 18
SS, *See* Suspended solids
St. Clair River Tunnel, 30
St. Stephen's Clock Tower, *See* Big Ben
Staffing, project, 62–64
State Environmental Quality Review Act, 2–3
State Historic Preservation Office, 121, 137
State Preservation award, 150
State Street Project, 136, 151–153, 155
  environmental review, 152
  mini-museum, 153
  regulations dilemma, 152
State-of-the-shelf technology, 196
Steel sheet piling, 42, 51, 54
Stern, Henry, 100
Stockholm Conference, 246
Subway construction, open cut, 81
Sunken tube tunnel construction, 32–33, 35
Surles, Richard, 98
Suspended solids, 71
Sweden, 82
Sykes, Dickie, 238

T

Tangent piles, 52
Taylor, Marilyn Jordan, 237
TBM, *See* Tunnel Boring Machine
Tennessee Saturn plant, 222
Thames river, 8, 11
Third World Countries, 82
Timber shoring, 42
Time-displacement plots, 15, 16
Times Square, 235
Tishman Construction Company, 133, 231, 240
Tree establishment techniques, 175, 176, 180
Tree pit, dysfunctional design, 171–172
  Belgian blocks, 172
  grates, detrimental, 171, 172
  layers, 171
  macro tree pit, 173
  porosity, 171
  roots, escaping, 171
  strategies and alternative systems, 172
  unit pavers, 172
Tree rooting zone, 159, 173, 182, 185, 186
  compaction, 173
  Dripline methods, 185
  load-bearing, 160
  media, 159
  protection zone, designating, 183–184, 187–188
  root plate, 185
  root-zone overlap, 170
  soil selection, 159
  Trunk formulae, 185
Trees and urban construction
  below-ground limitations, 159
  cantilevered pavements, 172–173
  collective forest canopy, 158, 182
  compaction control, 172–173
  container-cantilever systems, combination, 173
  containerized plants and planting beds, 170, 180
  containerized systems, 167
  failure to thrive, 159
  heat island, 158
  irrigation and fertilization systems, 158–159
  light-weight soil textural amendments, 169
  protecting, 180
  rooftop planting, 169
  species and cultivar selection, 158
  values and benefits, 158
  water use and soil volume, 170
Trees, media specification, 160–162
  design considerations, 167–169
  drainage design, 162
  plant survival, 160–162
  topsoil specifications, 161–162, 162
Tremie concrete seal, 52
Trump, Donald, 235
Tubes a manchette, *See* Compensation grouting
Tunnel Boring Machine, 64. *See also* Tunneling in cities
Tunnel construction, major challenges in Central London, 7
  archeology, 12–13
  history, 8
  London clay, advantages, 8
  material supply and disposal, 22, 24
  population, 8
Tunnel construction, parliamentary effect, 10–11
Tunnel worker's labor union, 61
Tunneling
  drilling and blasting, 65
  elevated structures, *versus*, 82
  environmental impacts, 64–65
  equipment failure, 65
  in cities, 64–66
  neighborhood impact, 64
  scheduling, 65
Tunnels, suitable locations in London, 11

U

U District, 241
U.S. Army Corps of Engineers, 71
U.S. Census Bureau, 233
U.S. Department of Energy and Environmental Protection Agency, 196, 209
U.S. Green Building Council, 196
U.S. Labor Department Compliance Officer, 63
U.S. Small Business Administration, 239
Underpinning challenges, 43–46
  brine system, 45
  needle beams, 43
  pit-underpinning, 43
  soil freezing, 46
United Nations Conference on the Human Environment, 246
University Book Store, 241
University District, 243
University District Farmers Market, 244
University of California at Berkeley, 237

University of Washington, 241
  construction controversy, 242
  Medical Center, 243
  School of Architecture, 244
  School of Communication, 244
University Park, 243
University Way Planning Group, 244
Urban, 171
Urban design and community relations, 93
Urban Horticulture Center, 244
USGBC, *See* U.S. Green Building Council
Utilities, considerations during construction, 79
UW, *See* University of Washington

V

VAV boxes, 85
Vegetation strips, 164–166
  drains, 166
  setback ordinances, 165
  utility corridors, 164
Victoria Line, 19
Vigilante, Theresa, 236
Vincent G. Kling & Associates, 84
VOCs, *See* Volatile organic chemical compounds
Volatile organic chemical compounds, 199, 201

W

W.A.N.E. Tree Systems, 173
Wakefield Sheeting, 51
Waldorf Astoria Hotel, 149
Walla Walla Road, 242
Ward's Island, *See* New York State Psychiatric Institute
Washington Heights, 95, 97, 99. *See also* New York State Psychiatric Institute
Washington, D.C., 171
Waste prevention and management, 201
  recycled content and renewable resources, 203
  strategies, 202–203
Wastewater treatment facilities, 69–70
  double-decking, 71, 72
  historical perspective, 70–71
  short period aeration process, 71
Water Street project, 136, 144, 150
  site preparation work, 148
  sunken ship, 145
Waterloo International, 11
Waterloo Station, 7, 10, 14, 24
WBE, *See* Women's Business Enterprises
Weiss, Lois, 235, 240
West Side Harlem, 72
Westminster Abby, 11
Westminster Bridge, 11
Westminster Station, 10, 11, 14
  spoil disposal, 22
Westminster, Palace, 11
Whisler, Patty, 244
William H. Gates School of Law, 241
Women's Business Enterprises, 232, 233, 234, 238, 239, 240
Women-owned business, 233
  construction, fastest growth, 231, 233
  cracking the concrete ceiling, 234
  engineering and architecture, 236, 237
  funding obstacles, 235
  real estate development, 235
Worldwatch Institute, 207